Quantum Minds: Exploring Quantum Computing and AI Synergy

Madhusudan Singh · Bharat S. Rawal

Quantum Minds: Exploring Quantum Computing and AI Synergy

From Qubits to Neural Nets—A Journey into Intelligent Computation

Springer

Madhusudan Singh
Department of Computer Science
and Engineering
The Pennsylvania State University
University Park, PA, USA

Bharat S. Rawal
Department of Computer Science
and Digital Technologies
Grambling State University
Grambling, LA, USA

ISBN 978-981-95-2497-6 ISBN 978-981-95-2498-3 (eBook)
https://doi.org/10.1007/978-981-95-2498-3

This Springer imprint is published by the registered company Springer Nature Singapore Pte Ltd.
The registered company address is: 152 Beach Road, #21-01/04 Gateway East, Singapore 189721, Singapore

Preface

The world stands at the dawn of a new era in computing. Quantum information science, once confined to speculative research papers, is rapidly crossing the threshold into the mainstream. Powerful quantum computers capable of tackling real-world problems are transitioning from science fiction fantasy into engineering reality.

Parallel breakthroughs in artificial intelligence, fueled by advances in data, algorithms, and specialized hardware like GPUs, have enabled unprecedented predictive capabilities for everything from medical diagnosis to autonomous vehicles. However, mounting evidence suggests that even the most potent AI today stays confined to a narrow intelligence sphere.

Quantum computing and AI convergence represent the next change in thinking that will shatter these limitations. Quantum machine learning can tap into radical new realms of computational power unlocked by quantum bits. Our classically constrained notions of data, training models, optimization, and simulation face disruption.

This book is an introductory guide to this emerging quantum AI landscape for forward-thinking professionals and students in the sciences, engineering, and business management. It demystifies the foundational concepts around quantum information processing and quantum-enhanced machine learning. A survey of near-term algorithms, applications, and enabling software/hardware ecosystems provides insight into practical quantum AI trends today.

We guess how quantum AI may transform organizational decision-making, planning, and competitive strategy as these capabilities mature. While forecasts are still uncertain, it is indisputable that the quantum computing revolution could engender a metamorphosis across industries on a scale rivaling the emergence of classical computing itself just decades prior.

Equipping technology and business leaders with literacy in quantum AI is vital to navigating this looming change in basic assumptions. We hope this book provides a valuable starting point for entering quantum artificial intelligence's weird yet tremendous world.

University Park, USA Madhusudan Singh
Grambling, USA Bharat S. Rawal

Contents

About the Authors

Madhusudan Singh Ph.D., SMIEEE, is an associate teaching professor and leads Blockchain Data Intelligence (Blockchain Innovation) Lab in the Department of Computer Science and Engineering at the College of Engineering at The Pennsylvania (Penn) State University, University Park, State College, PA, USA. He previously served as an associate professor and chair of Data Analytics in the Department of Entrepreneurship and led the Center for Blockchain Technology and Data Analytics at Long Island University, Brooklyn, New York. Before that, he was an assistant professor in the Business department and founder of the Quantum Computing Innovation Center at the Oregon Institute of Technology (OregonTech), USA; his previous roles include an assistant professor and led the Nexus Cybersecurity Research Center (Applied AI/Data Science and Blockchain Technology) at Woosong University in South Korea.

Bharat S. Rawal is a full Professor and a department head of Computer Science and Digital Technologies Department at Grambling State University, USA, and a senior fellow at US, Navy. His research focuses on network security, cloud computing, and security, blockchain, Big Data, analytical modeling, smart grid, and next-generation cyber defense. He has published several peer-reviewed conference proceedings and journal publications and has two patents. He is a member of (ACM, IEEE, IET, etc.). He received his D.Sc. degree in IT from Towson University Maryland. Previous Faculty Positions: Full Professor, Department of Chair Capitol Technology University, Associate Professor at Gannon University, Pennsylvania State University, Duke University, Industry Served: Biochem Pharmaceutical Industries, Ltd. Bashundhara Pharmaceutical, Ltd. Coracias Advance Technologies, LLC. Roles he served in the industry: Chairman, President, CEO, and Director of Marketing. His research interests are in quantum computing, post-quantum cryptography, cyber security, artificial intelligence, etc.

Part I
Foundations of Quantum Computing

Quantum computing stands for a radical departure from classical computing paradigms. It harnesses strange and counterintuitive quantum mechanical phenomena like superposition, entanglement, and quantum tunneling to perform computations in ways that conventional silicon chips cannot. To fully grasp quantum computing's transformative potential, management professionals must build foundational literacy on the quantum mechanical principles underlying this emerging technology.

This part provides a primer on quantum mechanics tailored for a non-physics audience. It aims to develop intuition on quantum phenomena and explain key quantum concepts using thought experiments and analogies accessible to business professionals without complex mathematical formalism. Understanding these elementary quantum principles will equip managers to visualize the workings of quantum systems better and appreciate their radically different information processing capabilities compared to classical computers.

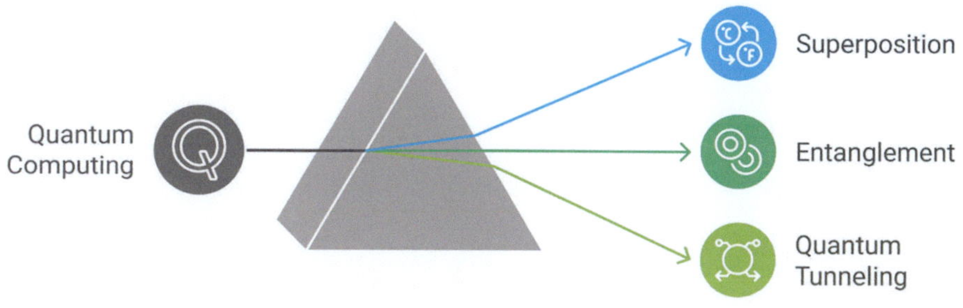

Fig. 1 Quantum computing overview

Chapter 1
Quantum Mechanics Primer for Non-physicists

1.1 Classical Versus Quantum Worldviews

Classical physics, developed by luminaries like Newton, Maxwell, and Einstein, is based on an intuitively logical, deterministic worldview. Objects exist in a singular, definite state at any moment. Take a coin—it can only be heads or tails. Measurements reveal pre-existing properties. Computational models in classical physics are based on bits existing as 0 or 1 [1] (Fig. 1.1).

Quantum mechanics fundamentally challenges the simplistic notion of reality, inviting you to explore a new and intriguing realm. In the quantum realm of atoms and subatomic particles, objects can exist simultaneously in an ambiguous superposition of multiple states and locations. A quantum bit or qubit can be 0 and 1 simultaneously. Instead of definite properties, objects only have probability and tendencies. Measurement does not reveal pre-existing values but instead causes a set of probabilities to collapse into a single observed value [2].

Two fundamental principles of quantum theory shatter classical assumptions—the uncertainty principle and the observer effect. The uncertainty principle tells that specific pairs of properties like position/momentum or time/energy cannot be precisely measured together. Observing one leads to fundamental uncertainty in the other. The observer effect, however, goes a step further. It notes that measurement doesn't just passively reveal pre-existing values. Instead, it actively influences the observed system (Fig. 1.2).

These bizarre quantum effects may seem counterintuitive from our everyday macroscopic perspective. However, numerous experiments have empirically validated that they accurately describe the behavior of the microscopic quantum world [3]. Understanding concepts like uncertainty, superposition, and entanglement is not just interesting, but it's also critical. It's the key to unlocking the immense power of quantum computing.

M. Singh and B. S. Rawal, *Quantum Minds: Exploring Quantum Computing and AI Synergy*, https://doi.org/10.1007/978-981-95-2498-3_1

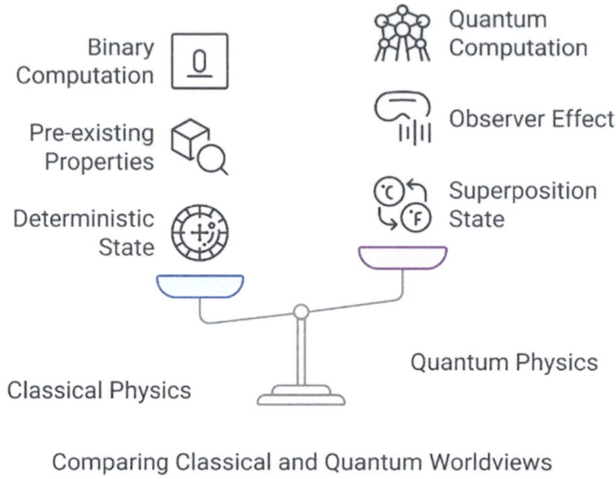

Fig. 1.1 Classical and quantum computing comparisons

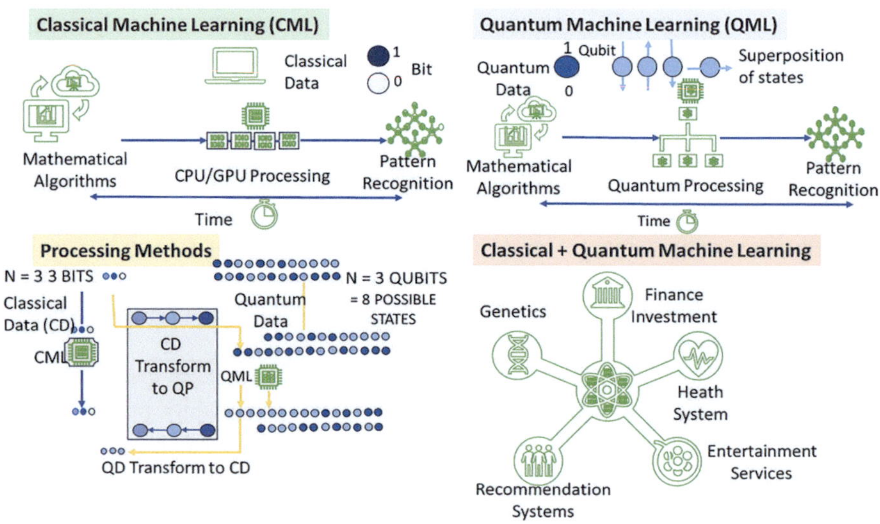

Fig. 1.2 Classical and quantum machine learning

1.2 Exploring Quantum States and Superposition

Qubits exist in a quantum superposition of 0 and 1, representing both values simultaneously. This enables a radically new information processing paradigm compared to classical bits holding singular values of 0 or 1 (Fig. 1.3).

To understand superposition, imagine a coin flipping in the air, caught at the moment it is heads-up and tails-up [4]. Quantum objects like electrons can exist

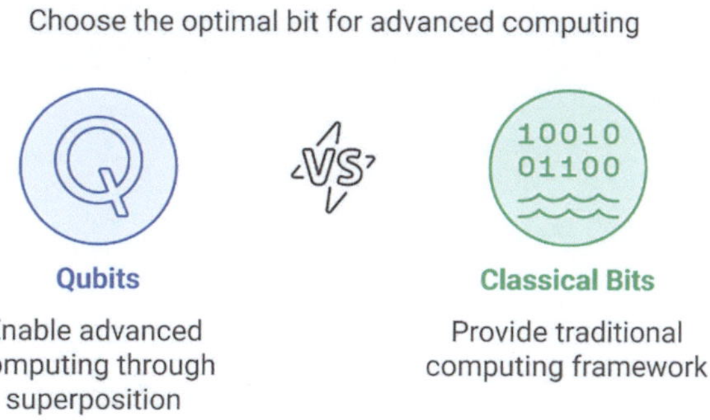

Fig. 1.3 Qubits versus classical bits

in such ambiguous mixed states. This thought experiment also reveals a critical difference between quantum and classical uncertainty. Classical physics says a coin mid-flip is either a head or a tail, but we don't know which one. Quantum mechanics says the coin is simultaneously both heads AND tails until it is measured.

In quantum computing hardware, qubits are realized using tiny, engineered objects like electrons, ions, or photons. Their quantum state is manipulated using laser pulses, microwave signals, or magnetic fields [5]. Observing or measuring a qubit causes its superposition to randomly collapse into a classical value of 0 or 1 based on underlying probability amplitudes. Repeated measurement of the same qubit yields different outcomes reflecting this inherent uncertainty (Fig. 1.4).

Quantum computers can exponentially scale information processing capabilities compared to classical bits by combining many qubits into large entangled systems [6]. While today's quantum processors use with 10–100 qubits, the goal is to engineer systems with millions of qubits eventually. Even a modest system of 300 qubits could store more values than atoms in the observable universe!

1.3 Quantum Entanglement and Its Peculiarities

Entanglement is described as the strangest and most counterintuitive phenomenon in quantum mechanics. When two qubits become entangled, their states get inextricably linked. Measuring one qubit instantaneously affects the other, even if large distances physically separate them.

This bizarre behavior violates conventional notions of cause-and-effect and locality in physics. Einstein dismissed entanglement as "spooky action at a distance" [7]. How can measuring qubit A here on Earth instantly influence entangled qubit B on Mars without clear communication? While philosophically troubling, entanglement is empirically confirmed through many experiments and forms the bedrock for quantum computing and cryptography (Fig. 1.5).

To illustrate entanglement, imagine two magical coins that always land on the same side when flipped and measured, even if separated across distant galaxies! This does not mean they show the same side all the time. Instead, measuring one coin

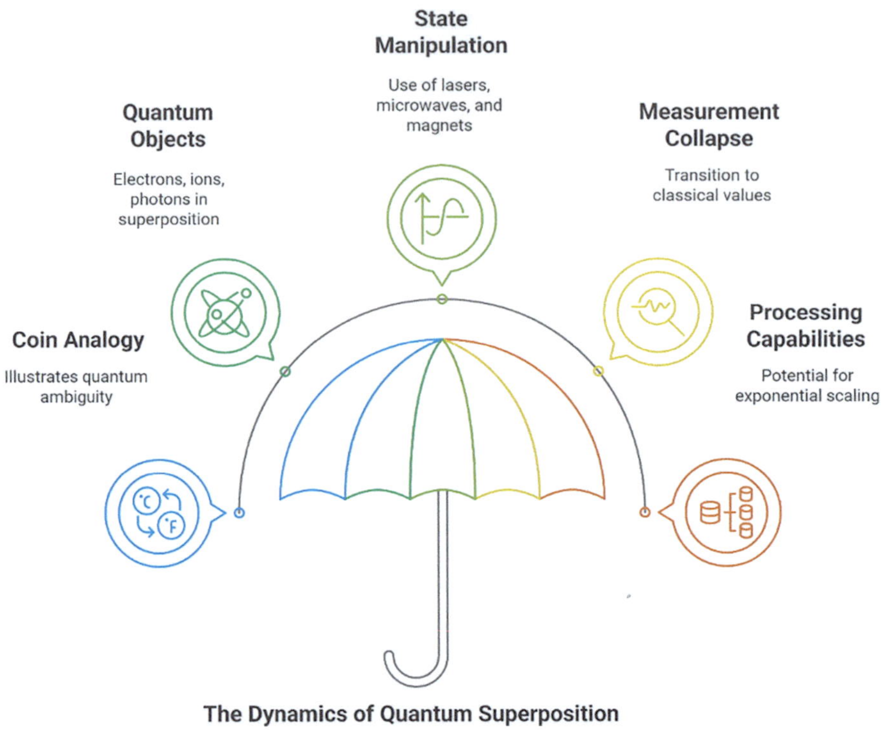

Fig. 1.4 Quantum superposition computing

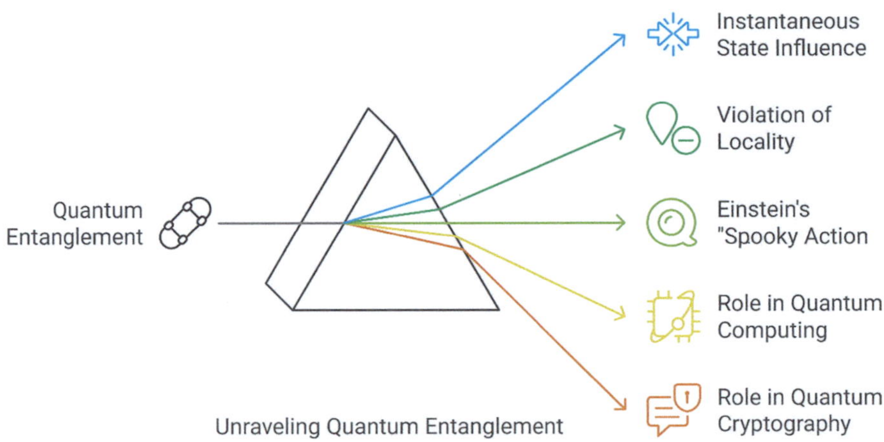

Fig. 1.5 Classical and quantum machine learning

instantaneously transmits information to the other coin about what side to show. Quantum entanglement works in similar bizarre ways [8].

Entanglement enables exponentially increasing quantum computers' information processing ability by linking qubits. A system of just 300 fully entangled qubits could represent more states than atoms in the universe [7]. By cleverly manipulating and measuring these entangled superpositions, quantum algorithms can rapidly solve problems considered intractable for classical computers (Fig. 1.6).

1.4 Quantum Gates and Quantum Circuits

Classical computers perform operations using logic gates like AND, OR, NOT that manipulate bits. Quantum computations also sequence logical operations on qubits using quantum logic gates [9]. These gates manipulate qubits in their ambiguous superposition states to enact desired transformations (Fig. 1.7).

Standard single-qubit gates include [10]:

- *Hadamard Gate (H)*—Puts qubit into superposition of 0 and 1.
- *Pauli-X Gate*—Flips or inverts the qubit state.
- *Phase Shift Gates*—Rotates the phase of qubit superposition.
- There are also two-qubit gates that entangle pairs of qubits:
  - *Controlled NOT (CNOT)*—Flips target qubit conditional on control qubit state.
  - *SWAP Gate*—Exchanges states between two qubits.

Sequencing quantum gates into circuits allows the enacting of multi-step algorithms on qubits. Analogous to classical Boolean logic circuits, quantum circuits transform input states into desired output states by chained operations. Carefully designed sequences of gates coupled with the measurement at the end can implement practical quantum algorithms like Shor's factoring algorithm, Grover's search, quantum Fourier transforms, etc.

Quantum circuit diagrams provide a shorthand for visualizing the step-by-step application of gates on qubits, from input states to final output. Understanding quantum circuits helps build intuition about how quantum programs manipulate quantum states (Fig. 1.8).

Quantum computing platforms like IBM Quantum Experience enable users to construct quantum circuits combining gates of their choice using simple drag-and-drop interfaces. Executing the circuit on actual quantum hardware shows the strange effects of superposition, entanglement, and measurement. As quantum processors scale up, more complex circuits unlocking significant business value will become workable.

In short, Quantum computing is based on deeply counterintuitive principles of quantum mechanics, such as uncertainty, superposition, and entanglement. It builds intuition about these concepts using thought experiments and analogies accessible to non-physicists. Understanding ambiguous quantum states, spooky entanglement,

Fig. 1.6 Quantum entanglement computing

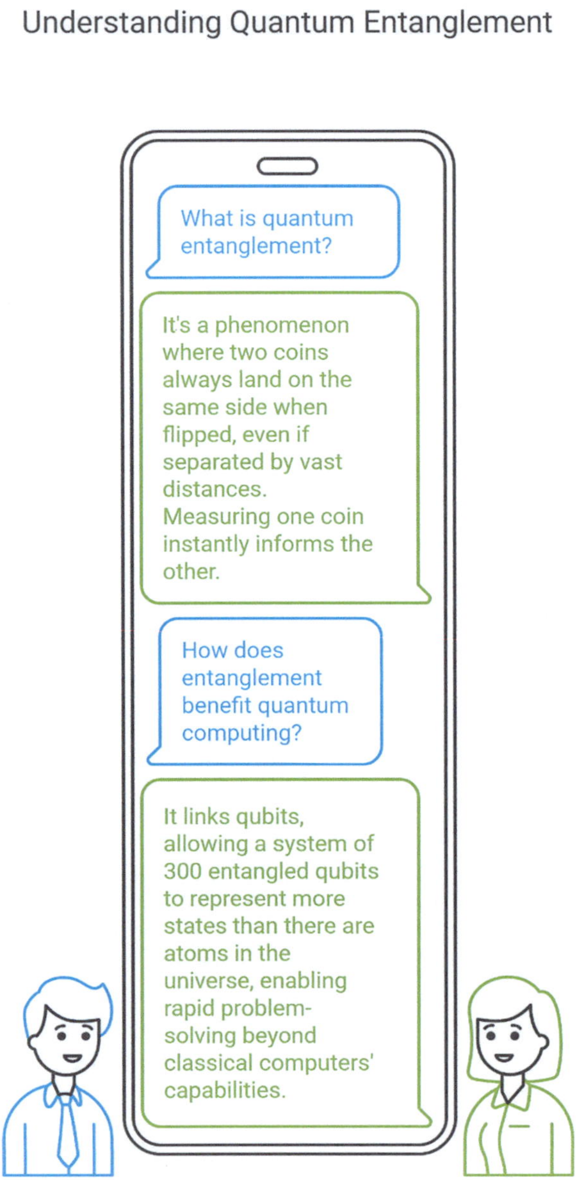

and principles of quantum circuits will equip management professionals to better grasp the workings and potential of quantum information systems. It's shown in Fig. 1.9.

However, quantum mechanics limits our ability to visualize and model quantum processes. As physicist Richard Feynman noted, "If you think you understand quantum mechanics, you don't." Developing a basic mental model of peculiar

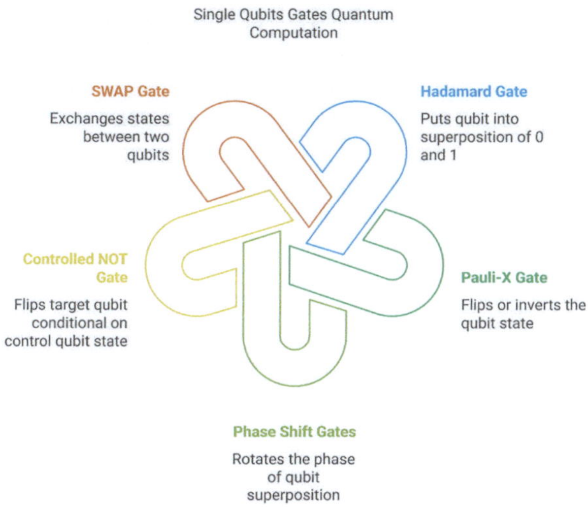

Fig. 1.7 Single-qubit gates' quantum computation

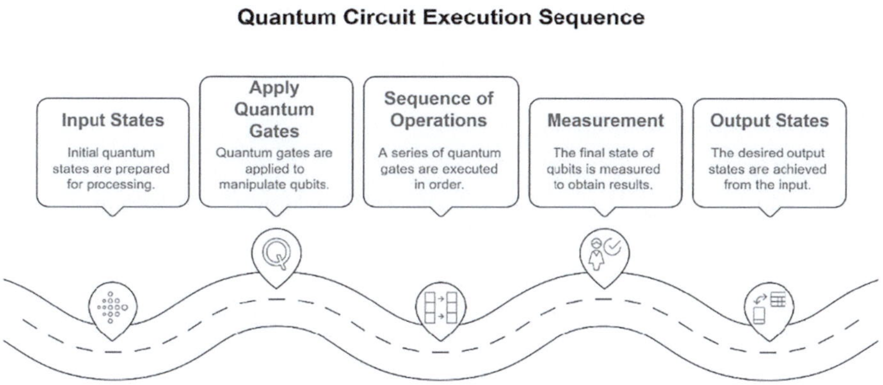

Fig. 1.8 Quantum circuit execution

quantum phenomena like superposition, interference, tunneling, and entanglement will prepare business leaders to begin exploring quantum computing applications for everything from risk analysis to supply chain optimization [11].

1.5 Building Blocks of Quantum Computing

While classical computers encode information in binary bits, such as 0 or 1, quantum computers use quantum bits or qubits, which can exist in superpositions of 0 and 1. Qubits enable a radically new information processing paradigm compared to classical

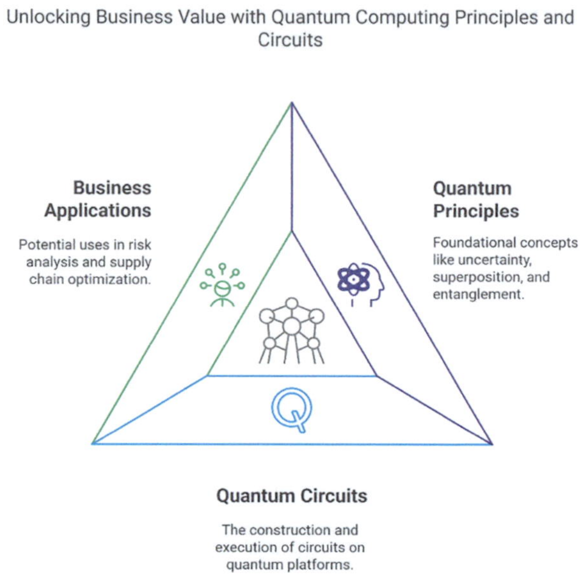

Fig. 1.9 Quantum computing principles and circuits

bits as shown in Fig. 1.10. This section overviews qubits as the primary information units in quantum systems and quantum gates and circuits, which operate on these qubits to run algorithms [12].

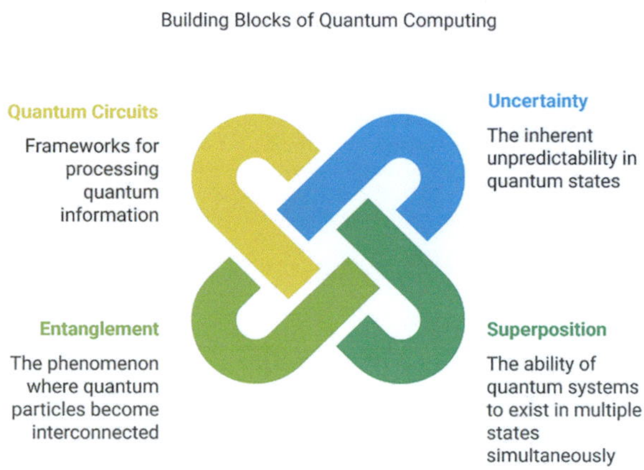

Fig. 1.10 Quantum computing blocks

1.6 Qubits: The Quantum Counterpart to Classical Bits

A qubit or quantum bit is a quantum computer's fundamental unit of information analogous to the classical bit used in regular computers. However, while a classical bit must be either 0 or 1 at any point, a qubit can simultaneously exist in a superposition of both states (Fig. 1.11).

This ability to encode 0 and 1 simultaneously directly results from quantum mechanical phenomena and enables qubits to process information in ways not possible with simple binary digits. The information processing ability of a quantum system scales exponentially with the number of qubits, unlike classical systems, which scale linearly. However, this power comes with a catch—qubit states are fragile, and any external disturbance makes them decay into classical 0/1 states through decoherence. State fragility is still the most significant engineering challenge in constructing practical qubit devices.

Physically, a qubit can be represented by any microscopic quantum system with two discrete quantum states that can be manipulated using external controls like lasers, magnetic fields, or microwave signals (Fig. 1.12).

Various qubit modalities have been developed:

- **Superconducting qubits**—tiny circuits of Josephson junctions cooled to near absolute zero. Examples are used in IBM, Google, and Righetti quantum computers.
- **Trapped ion qubits**—Individual atoms suspended by lasers and manipulated using laser pulses and electromagnetic fields and used by IonQ.
- **Photon qubits**—Encoding qubit states in properties of a photon, such as polarization or path, and being studied for long-range quantum communication.
- **Topological qubits**—More stable states encoded properties of exotic materials. It is still in the early stage of research.

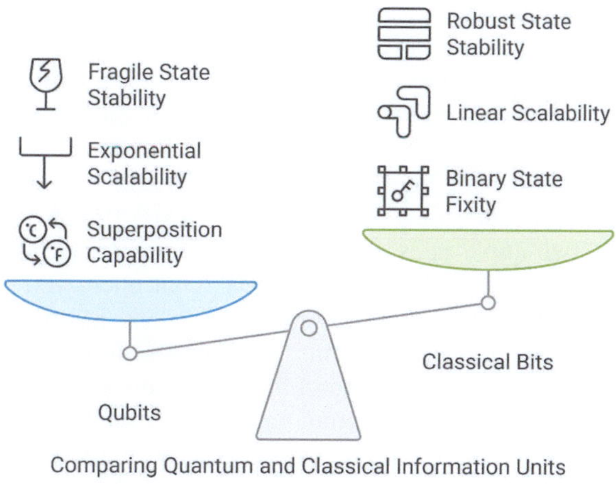

Fig. 1.11 Qubits: the quantum computing counterpart to classical bits

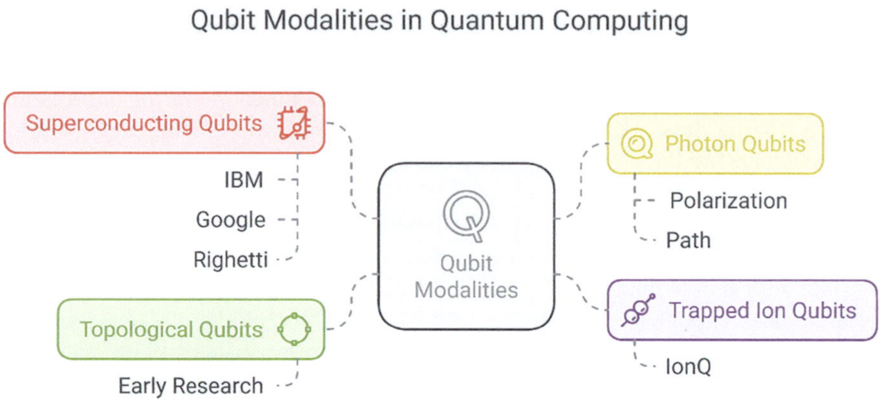

Fig. 1.12 Qubits modalities in quantum computing

Regardless of physical implementation, all qubits share the same underlying quantum information processing capabilities and can be described mathematically in the same formalism. Operating systems and algorithms do not depend on the specific qubit technology. Ongoing research aims to build qubit devices that balance stability, scalability, controllability, and integration into complete quantum computing systems.

1.7 Quantum Gates and Operators: Manipulating Qubits

To perform practical computations, qubits must be precisely initialized, controlled, transformed, and measured through careful sequences of quantum operations. Single-qubit operations manipulate individual qubits, while multiple qubit operations allow interactions between qubits to create quantum entanglement. Together, these operations form the essential toolset for enacting quantum algorithms.

The elementary operations on qubits are represented by quantum gates—the quantum analogs of logic gates in a classical computer. Single-qubit gates like Hadamard, Pauli-X, and Z gates enact fundamental transformations by altering the probability amplitudes of the qubit's 0 and 1 states. Other important single-qubit gates include phase shift and rotation gates (Fig. 1.13).

Two-qubit gates entangle pairs of qubits by correlating their states. The most common two-qubit gate is CNOT (controlled-NOT) [13], which flips the second qubit conditional on the first qubit's state. Multi-qubit gates enact entanglement across more than two qubits. Special measurement operations are also used to read out qubit states at the end of the computation (Fig. 1.14).

Together, these quantum gates form a universal gate set—any arbitrary unitary transformation on qubit states can be decomposed into sequences of these fundamental gates. Like the universality of NAND gates in classical digital logic, the

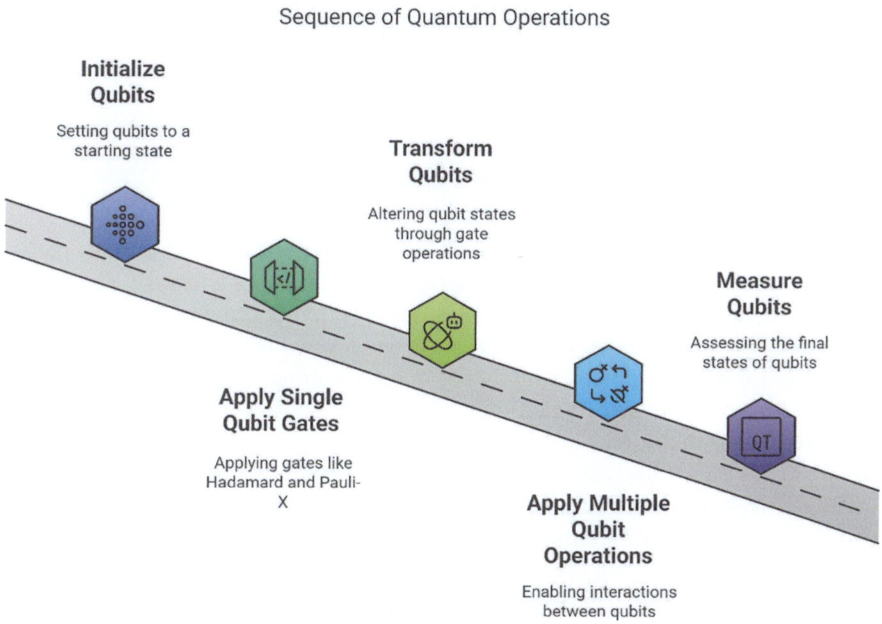

Fig. 1.13 Quantum operations sequence

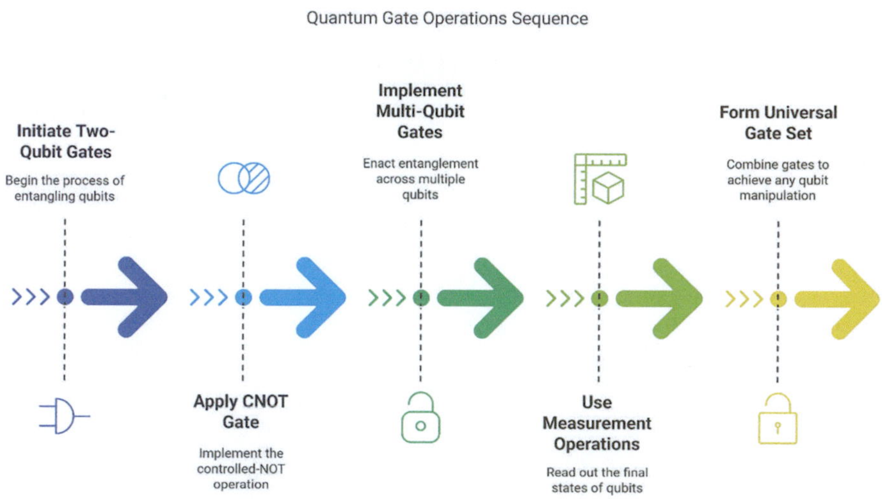

Fig. 1.14 Quantum operations sequence

quantum gate library allows any possible qubit state manipulation required for an algorithm through circuit synthesis.

1.8 Quantum Circuits: Assembling Quantum Operations

A quantum circuit diagrams the sequence of quantum operations needed for a particular algorithm by representing gates as symbols and qubits as wires. By abstracting away the underlying math and physics, quantum circuits provide a shorthand for understanding complex quantum programs at an elevated level.

Quantum circuits begin by initializing qubits to prepared input states, enact a sequenced cascade of gate operations, leverage entanglement to enable massive parallelism, and finally read out qubit states through measurement. The output of this circuit stands for the result of the quantum computation [14] (Fig. 1.15).

The weird properties of qubits and quantum physics enable circuits with just a few hundred qubits and gates to implement computations that would take conventional supercomputers longer than the universe's age! However, quantum states are fragile and susceptible to errors, noise, and decoherence. Quantum error correction and fault-tolerance techniques will be critical to scale quantum circuits to large enough sizes where they can deliver practical business advantages over classical systems.

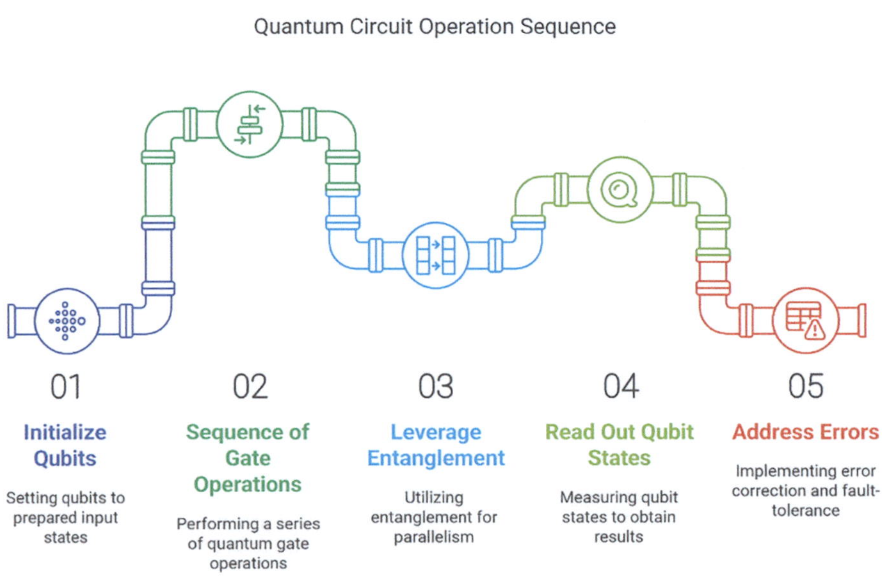

Fig. 1.15 Quantum circuit operation sequence

Summary

Qubits, quantum gates, and quantum circuits form the basic building blocks of quantum information processing systems and equip them with radically different computational powers than classical computers. Mastering these elementary concepts will help managers understand how quantum control techniques manipulate qubit superpositions and entanglement to run advanced quantum algorithms and applications not feasible on conventional machines.

References

1. Nielsen, M. A., & Chuang, I. L. (2010). *Quantum computation and quantum information* (10th anniversary ed.). Cambridge University Press.
2. Preskill, J. (2018). Quantum computing in the NISQ era and beyond. *Quantum, 2*, 79–95.
3. Cox, B., & Forshaw, J. (2011). *The quantum universe: Everything that can happen does happen.* Allen Lane.
4. Griffiths, D. J., & Schroeter, D. F. (2018). *Introduction to quantum mechanics* (3rd ed.). Cambridge University Press.
5. Blood, C. (2010). *A primer on quantum mechanics and its interpretations.* arXiv.
6. Bengtsson, I., & Życzkowski, K. (2017). *Geometry of quantum states: An introduction to quantum entanglement* (2nd ed.). Cambridge University Press.
7. DiVincenzo, D. P. (1997). *Quantum gates and circuits.* arXiv:quant-ph/9705009.
8. Guillaud, X., & Pinkse, P. (2023). Detecting entanglement-generating circuits in cloud-based quantum computing services. *iComputing*, *5*(1), 51.
9. Benchmarking Quantum Gates and Circuits. (2025). *Chemical Reviews*, *125*(9), 6105–6150.
10. Blais, A., Grimsmo, Å. L., Girvin, S. M., & Wallraff, A. (2020). Circuit quantum electrodynamics. arXiv:2005.12667.
11. Krantz, P., et al. (2019). A quantum engineer's guide to superconducting qubits. arXiv:1904.06560.
12. Xiang, Z.-L., et al. (2012). Hybrid quantum circuits: superconducting circuits interacting with other quantum systems. arXiv:1204.2137.
13. Krantz, P., et al. (2019). *A quantum engineer's guide to superconducting qubits.*
14. High-fidelity parallel entangling gates on neutral-atom qubits. (2023). *Nature*, *622*, 820–825.

Chapter 2
Overview of Quantum Computing

Quantum computing is a revolutionary new model of computation based on principles of quantum physics. This chapter will provide a brief history of quantum theory, explain fundamental quantum concepts like qubits and entanglement, compare quantum systems to classical systems, and survey some quantum algorithms and applications.

2.1 Brief History of Quantum Physics

The origins of quantum theory trace back to 1900 when physicist Max Planck first introduced the idea of energy quanta to describe the frequency of radiation emitted by heated objects, known as blackbody radiation [1]. This broke from the classical physics view of electromagnetic waves to explain experimental observations (Fig. 2.1).

In 1905, Albert Einstein extended Planck's quantized energy model to explain the photoelectric effect, proposing that light is absorbed and emitted in discrete packets called photons [2]. This contributed to the emerging field of quantum theory, which shows that energy and matter exhibit particles and wave properties at microscopic scales.

1923, Louis de Broglie proposed that any particle has an associated wavelength [3]. This wave-particle duality means quantum particles like electrons can behave like waves and vice versa. Werner Heisenberg advanced quantum mechanics in 1925 with the famous uncertainty principle, which imposed intrinsic probabilistic limits on measuring properties like position and momentum [4].

M. Singh and B. S. Rawal, *Quantum Minds: Exploring Quantum Computing and AI Synergy*, https://doi.org/10.1007/978-981-95-2498-3_2

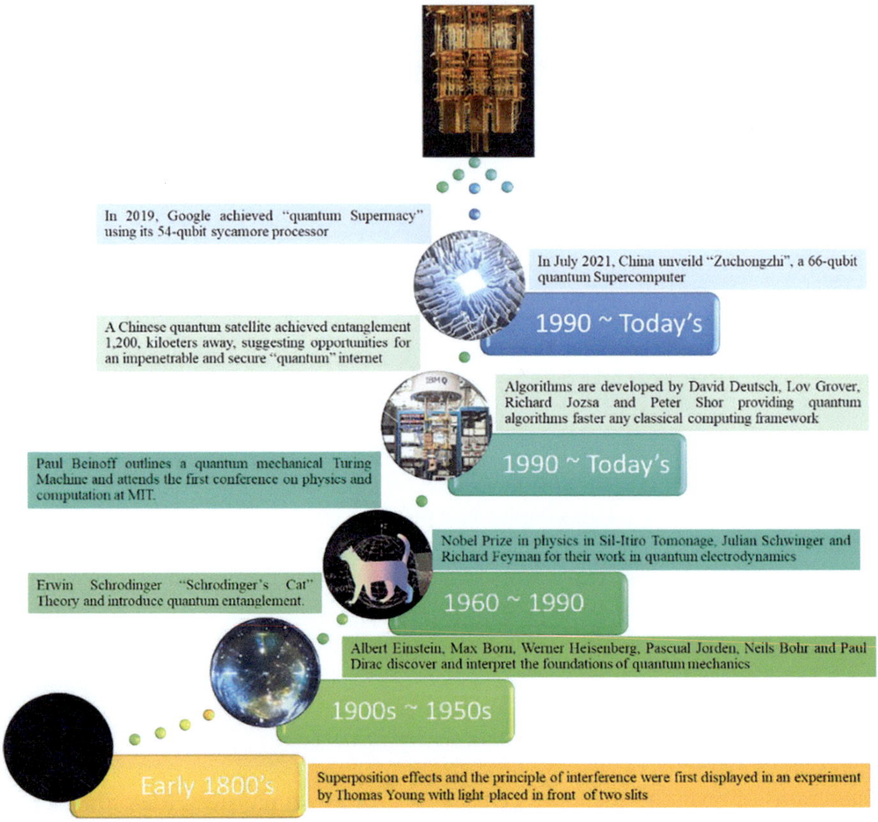

Fig. 2.1 Development periods of quantum computing

1926, Erwin Schrödinger formulated the fundamental quantum mechanical wave equation to model particle behavior [5]. This mathematical waveform underpins much quantum theory. Schrödinger showed that particles exist in a probabilistic superposition of states before measurement causes them to deterministically "collapse" into a classical state (Fig. 2.2).

These pivotal breakthroughs in the first few decades of the twentieth century established the foundations of quantum theory and radically transformed our understanding of physics at microscopic scales. The concepts of quantization, wave-particle duality, uncertainty, and quantum superposition contradicted prevailing classical views of reality—leading Richard Feynman to state that "nobody understands quantum mechanics famously."

Quantum principles defy human intuition, but the theory has been repeatedly verified experimentally and enables astonishing technological applications, from transistors to lasers. Quantum computing aims to harness these quantum effects as a new mode of information processing.

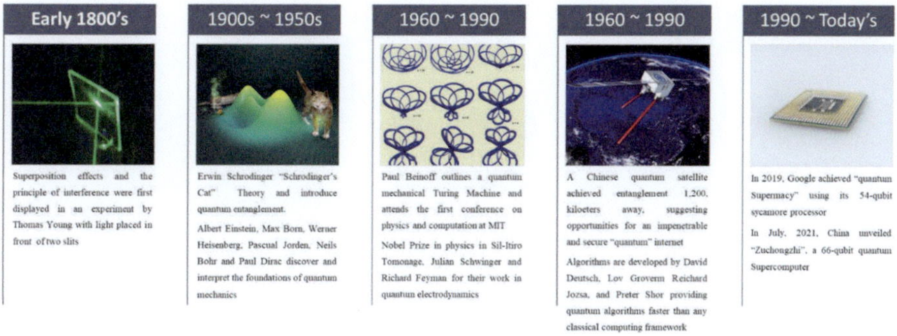

Fig. 2.2 Timeline of critical developments in quantum physics

2.2 Quantum Mechanics

"Quantum mechanics" is a fundamental physics theory that describes particles' behavior at the most minor scales, like atoms and subatomic particles. It's a bit different from the everyday world we experience and can seem strange at first.

Example Think about flipping a coin. When you toss it in the air, it's simultaneously in a state of both heads and tails. It's only when it lands that it becomes heads or tails. Quantum particles behave like a coin in the air, existing in multiple states until they are measured or observed.

Here's a basic breakdown of how quantum computing works:

1. *Qubits*: Instead of regular bits (0 or 1: the on/off switches in classical computers), quantum computers use qubits. Qubits can be both on and off simultaneously, which is called superposition.
2. *Superposition*: Imagine we can be in two places at once. In quantum computing, superposition allows qubits to perform multiple calculations simultaneously.
3. *Entanglement*: Qubits can be entangled, meaning they are connected so that the state of one qubit can instantly affect the state of another, no matter how far apart they are. When two qubits are intertwined, the state of one (whether it's 0 or 1) can change based on the state of the other. This makes it possible for qubits to link in ways that classical bits can't.
4. *Quantum Gates*: These are like circuits in classical computers but are used to perform operations on qubits by manipulating their quantum states. Quantum computers use quantum gates, like logic gates in traditional computers, to do math. Quantum gates change the states of qubits by using operations that follow the rules of quantum physics.
5. *Quantum Decoherence*: Keeping qubits in their quantum state requires frigid temperatures and careful control since the system can lose its quantum properties when interacting with its surroundings. This is called decoherence.

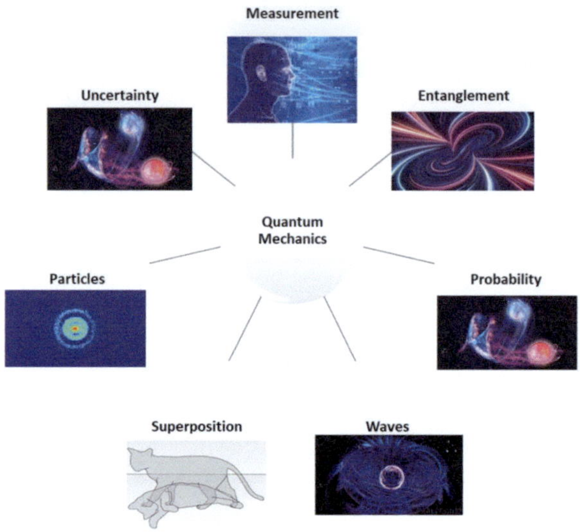

Fig. 2.3 Overview of quantum mechanics

Quantum Computing can provide faster problem-solving, improved security and encryption, scientific breakthroughs, better optimization, and more powerful artificial intelligence systems (Fig. 2.3).

2.3 Explanation of Qubits, Superposition, and Entanglement

Quantum computing is based on units of quantum information called quantum bits or "qubits." Whereas classical computers operate using bits represented as 0 or 1, qubits can exist in a quantum superposition of both 0 and 1 simultaneously before measurement [6].

Qubits employ quantum two-level systems as their physical implementation, like an electron's spin or a photon's polarization. Let $|0\rangle$ and $|1\rangle$ denote the two basis qubit states. Thanks to quantum superposition, a qubit $|\psi\rangle$ can exist in any linear combination of $|0\rangle$ and $|1\rangle$:

$$|\psi\rangle = \alpha|0\rangle + \beta|1\rangle$$

where α and β are probability amplitudes such that $|\propto|^2$ is the probability of observing $|0\rangle$ upon measurement, and $|\beta|^2$ is the probability of observing $|1\rangle$. The amplitudes are complex numbers that interfere in figuring out probabilities.

For example, an equal superposition state would be:

$$|\psi\rangle = \left(1/\sqrt{2}\right)|0\rangle + \left(1/\sqrt{2}\right)|1\rangle$$

Fig. 2.4 Showing qubit superposition

This means there is a 50% chance of measuring 0 and a 50% chance of measuring 1 in this qubit state. Superposition enables a qubit to represent both 0 and 1 simultaneously (Fig. 2.4).

Multiple qubits can also exhibit entanglement, where their states depend on each other regardless of physical distance. Measuring one qubit affects the other instantaneously per quantum mechanics. Einstein termed this "spooky action at a distance."

For example, a two-qubit entangled state:

$$|\psi\rangle = \left(1/\sqrt{2}\right)(|00\rangle + |11\rangle)$$

If qubit one is measured as 0, qubit two must measure 0. Entanglement enables nonlocal correlations between qubits (Fig. 2.5).

These phenomena have no equivalent in classical systems and are vital to providing quantum computers with their power.

2.4 Comparison of Classical Computing

Classical computers use bits with discrete 0 and 1 values that translate into binary digital states representing data and program instructions [7]. Operations use Boolean logic gates like AND, OR, and NOT to transform input bits into output results.

Fig. 2.5 Two qubit entanglement

Classical computation is deterministic, and bits exist in only one state at a time. N bits have 2N possible states, enabling exponential storage capacity. However, each operation acts on bits individually, so computers must perform program instructions sequentially.

Quantum computers harness quantum bits (qubits) and principles like superposition and entanglement to enable new modes of information processing. Some key differences include:

- Qubits can represent 0 and 1 simultaneously via superposition before measurement.
- Qubit states exhibit probabilistic rather than deterministic behaviors.
- Quantum parallelism allows simultaneously operating on all possible states of a multi-qubit register.
- Entanglement and interference enable nonlocal correlations between qubits.
- Reading out qubit results collapses superpositions into definite classical states.

While classical computers manipulate individual bits step-by-step, quantum computers can transform large-scale superposition states in parallel for specific workloads. This offers the potential for exponential speedups compared to classical algorithms (Figs. 2.6 and 2.7).

2.5 Overview of Quantum Algorithms and Applications

Researchers have discovered several quantum algorithms that offer asymptotic improvements over the best-known classical approaches by exploiting quantum mechanical effects [8]. Some prominent examples include:

- *Shor's Algorithm*—Integer factorization in polynomial time to break public-key encryption like RSA.
- *Grover's Algorithm*—Search databases quadratically faster than classical algorithms.

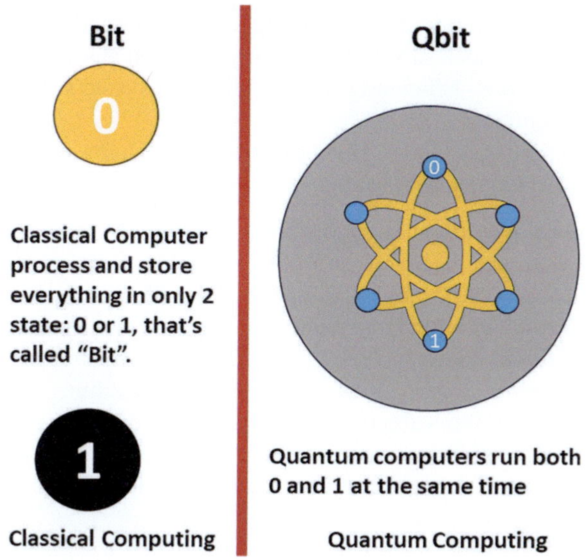

Fig. 2.6 Classical bits versus quantum qubits

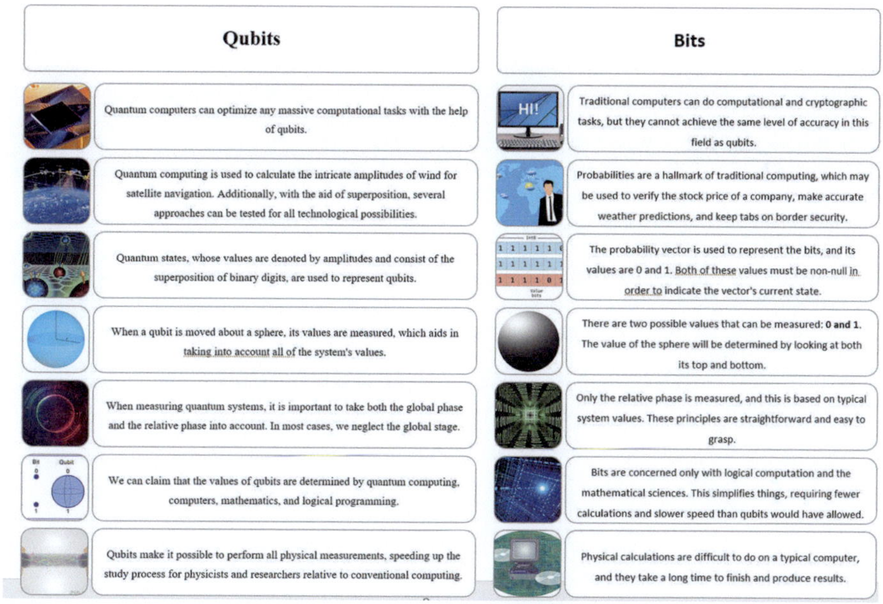

Fig. 2.7 Difference between classical bits versus quantum qubits

- *Quantum Fourier Transform*—Efficiently extracts periodicity from quantum states. They are used as a subroutine in many quantum algorithms.
- *Quantum Simulation*—Efficiently models quantum systems that are intractable on classical computers.
- *Quantum Machine Learning*—Quantum neural networks and kernel methods promise improved machine learning.

These algorithms may one day enable breakthroughs in cryptography, optimization, chemistry, medicine, finance, and other complex domains. However, major obstacles around qubit volume, error correction, decoherence, and hardware stability must first be overcome to build large-scale practical quantum computers.

Today's applications focus on areas like quantum chemistry, materials science, optimization, sampling applications, and quantum machine learning. Hybrid classical-quantum algorithms aim to achieve incremental advantages by using small quantum processors that are now accessible as co-processors to classical systems. Realizing the full potential of quantum computing remains a long-term challenge.

References

1. Planck, M. (1901). On the law of distribution of energy in the normal spectrum. *Annalen der Physik, 4*, 553.
2. Einstein, A. (1905). Concerning a heuristic point of view toward the emission and transformation of light. *Annalen der Physik, 17*, 132–148.
3. De Broglie, L. (1923). Waves and quanta. *Nature, 112*, 540.
4. Heisenberg, W. (1927). Über den anschaulichen Inhalt der quantentheoretischen Kinematik und Mechanik. *Zeitschrift für Physik, 43*, 172–198.
5. Schrödinger, E. (1926). An undulatory theory of the mechanics of atoms and molecules. *Physical Review, 28*, 1049–1070.
6. Nielsen, M., & Chuang, I. (2010). *Quantum computation and quantum information*. Cambridge University Press.
7. Von Neumann, J. (1993). First draft of a report on the EDVAC. *IEEE Annals of the History of Computing, 15*(4).
8. Montanaro, A. (2016). Quantum algorithms: an overview. *NPJ Quantum Information, 2*(1).

Chapter 3
Quantum Computing Keys

Quantum computing introduces several radical new concepts that contradict classical intuition yet are formally described by the mathematics of quantum theory. This chapter provides an in-depth explanation of the key theoretical ideas that enable the power of quantum computation, including the qubit, superposition, interference, entanglement, measurement, and quantum parallelism.

3.1 The Qubit

The quantum bit or qubit is the fundamental unit of quantum information [1]. Qubits extend the familiar classical bit into the quantum realm by leveraging two-level quantum systems. Whereas classical bits can only represent discrete 0 and 1 states, qubits exhibit quantum effects like superposition and entanglement that have no analog in classical systems.

Physically, qubits are implemented by quantum mechanical two-level systems. Examples include an electron's spin, a photon's polarization, the ground and excited states of an ion or artificial atom, or current states in a superconducting loop. Mathematically, the two orthogonal basis states of a qubit are denoted as: $|0\rangle$, $|1\rangle$.

These basis states are analogous to classical bits 0 and 1 states. However, thanks to quantum superposition, a qubit can also exist in any linear combination of $|0\rangle$ and $|1\rangle$:

$$|\psi\rangle = \propto |0\rangle + \beta|1\rangle$$

where α and β are complex probability amplitudes. The probability of measuring $|0\rangle$ is given by $|\alpha|^2$, and the probability of measuring $|1\rangle$ is given by $|\beta|^2$. The amplitudes obey a normalization constraint such that:

M. Singh and B. S. Rawal, *Quantum Minds: Exploring Quantum Computing and AI Synergy*, https://doi.org/10.1007/978-981-95-2498-3_3

$$|\alpha|^2 + |\beta|^2 = 1$$

As an example, consider the equal superposition state:

$$|\psi\rangle = \left(1/\sqrt{2}\right)|0\rangle + \left(1/\sqrt{2}\right)|0\rangle$$

There is a 50% probability of observing a 0 and a 50% chance of observing a one upon measurement. This ability for a qubit to exist in a superposition of 0 and 1 provides its power (Fig. 3.1).

When a measurement occurs, the qubit's superposed state collapses to a single classical value of 0 or 1 based on the probability amplitudes. Before measurement, the qubit effectively represents both 0 and 1 simultaneously.

3.2 Superposition

Superposition is the phenomenon where a quantum system exists in multiple states simultaneously [2]. This has no analog in classical physics and is counterintuitive. A simple example is Schrödinger's famous thought experiment, in which a cat can exist in a superposition of both alive and dead states inside a box until observed.

Qubits leverage quantum superposition within their two-dimensional state space. As shown earlier, a qubit $|\psi\rangle$, in general is described by:

$$|\psi\rangle = \alpha|0\rangle + \beta|1\rangle$$

The probabilities $|\alpha|^2$ and $|\beta|^2$ represent the weighting of the $|0\rangle$ and $|1\rangle$ states. However, each basis state is associated with its distinct physical state in the underlying qubit implementation. So before measurement, the qubit exists in both the $|0\rangle$ and $|1\rangle$ states simultaneously.

Unlike a classical bit, this enables a qubit to represent 0 and 1 simultaneously. Superposition is what gives quantum parallelism its power. The Bloch sphere provides a convenient geometric representation of a single qubit state space, as shown in Fig. 3.2.

All permissible qubit states correspond to points on the Bloch sphere. Superposition enables quantum states to exist in any complex combination, not just the classical basis states.

A qubit, which is the quantum counterpart of a classical bit, possesses a significantly larger number of potential states. The representation of states can be achieved by utilizing an arrow that indicates a certain geographical place on a sphere. In this context, it can be posited that the **north pole can be assigned a numerical value of 1**, whereas the **south pole can be assigned a numerical value of 0**. The additional sites exhibit quantum superpositions involving the states of 0 and 1.

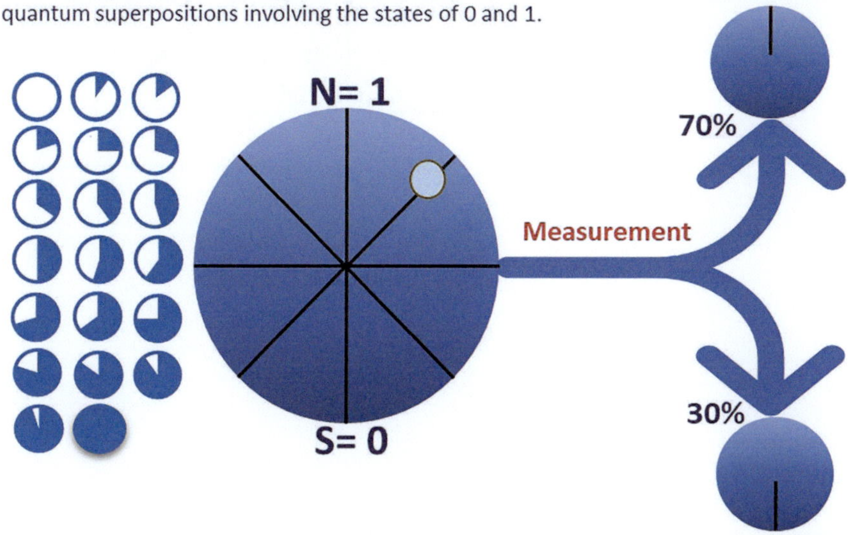

The QUBIT possesses the potential to hold a seemingly boundless quantity of information due to its ability to store an unlimited series of digits within its coordinates. However, the extraction of information from a qubit necessitates a measurement. According to the principles of quantum physics, when a qubit undergoes measurement, the outcome is invariably a classical bit, specifically either **a 0 or a 1**. The likelihood of each occurrence is contingent upon the **qubit's "latitude".**

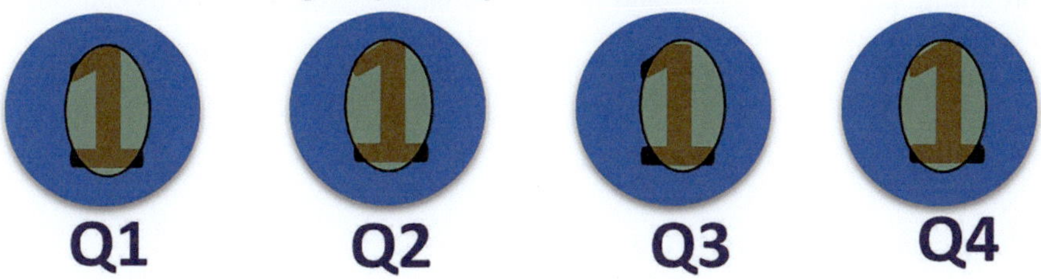

Fig. 3.1 Qubit superposition

3.3 Interference

Quantum interference arises when multiple quantum probability amplitudes interfere with each other, either constructively or destructively [3]. This causes non-classical effects observed in phenomena like the double slit experiment.

Consider a qubit in state $|\psi\rangle$ evolving in time under the dynamics of a unitary operator U:

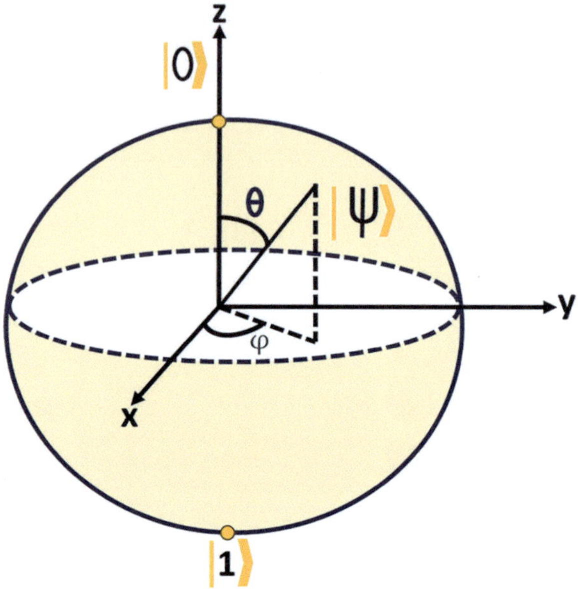

Fig. 3.2 Bloch sphere

$$|\psi\rangle = \alpha|0\rangle + \beta|1\rangle$$

$$U|\psi\rangle = \alpha'|0\rangle + \beta'|1\rangle$$

The probability amplitudes α and β can interfere, either increasing or decreasing the resulting probabilities $|\alpha|^{'2}$ and $|\beta|^{'2}$ based on phase factors. Qubits can cancel out or enhance each other.

For example, given a 50/50 superposition state:

$$|\psi\rangle = \left(1/\sqrt{2}\right)(|0\rangle + |1\rangle)$$

Under a unitary that flips $|0\rangle$ to $-|0\rangle$:

$$U|\psi\rangle = \left(1/\sqrt{2}\right)(-|0\rangle + |1\rangle)$$

The amplitudes destructively interfere, resulting in a 0% probability of measuring $|0\rangle$.

Quantum interference enables qubit amplitudes to interact in wavelike ways without classical explanation. Carefully designed interference is critical for many quantum algorithms.

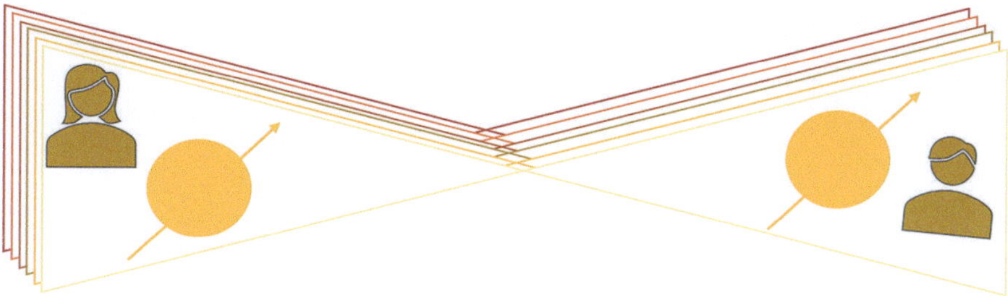

Fig. 3.3 Entanglement

3.4 Entanglement

Entanglement is a non-local quantum correlation between two or more quantum particles or systems [4]. Measurements on one entangled particle affect the outcomes of measurements on its partners, even when separated by large distances.

Entangled qubit pairs exhibit probabilistic correlations stronger than any possible classical explanation. For example, consider the Bell state describing a two-qubit system:

$$|\Phi+\rangle = \left(1/\sqrt{2}\right)(|00\rangle + |11\rangle)$$

If the first qubit is measured as $|0\rangle$, the second qubit must measure $|0\rangle$ also. However, before measurement, the state represents a superposition of both possibilities (Fig. 3.3).

Entanglement leads to surprising non-classical effects. It enables quantum teleportation, cryptography, dense coding, and other applications. Generating and maintaining qubit entanglement is an ongoing challenge in building quantum computers and networks.

3.5 Measurement

Measurement collapses a qubit's superposed state to a single classical value [5]. Consider a qubit in the state:

$$|\psi\rangle = \alpha|0\rangle + \beta|1\rangle$$

When measured in the $\{|0\rangle, |1\rangle\}$ basis, the probability of outcome 0 is given by $|\alpha|^2$ and outcome one by $|\beta|^2$. After measurement, the qubit's state collapses to just the observed eigenstate.

For example, given:

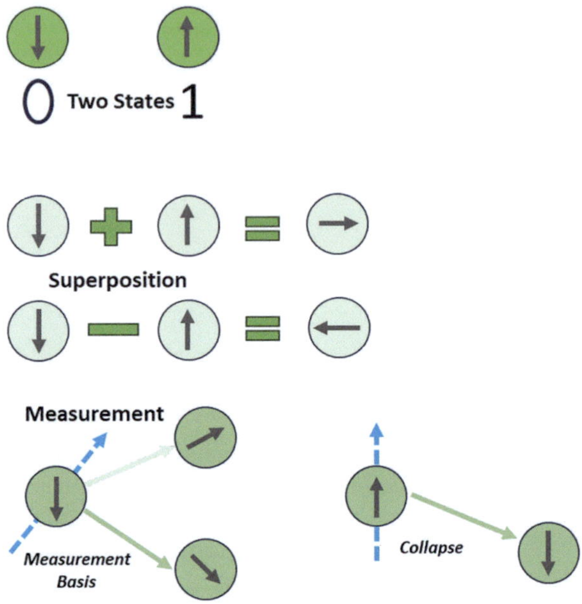

Fig. 3.4 Measurement

$$|\psi\rangle = \left(1/\sqrt{3}\right)|0\rangle + \left(\sqrt{2/3}\right)|1\rangle$$

There is a 1/3 probability of measuring 0, collapsing the state to:

$$|\psi\rangle = |0\rangle$$

And 2/3 probability of measuring 1, collapsing the state to:

$$|\psi\rangle = |1\rangle$$

Measurement irreversibly collapses the quantum superposition. Avoiding unintended measurement is critical in quantum circuits, and quantum operations must preserve superposition states (Fig. 3.4).

3.6 Quantum Parallelism

Quantum parallelism refers to the property where qubit superposition enables evaluating a function on all possible inputs simultaneously [6]. Consider a function $f(x)$ with N possible inputs x. Evaluating $f(x)$ classically would require N steps trying each x.

But if x is represented using $\log_2 N$ qubits, a superposition of all $2N$ input states can be created as $|\psi\rangle$. Applying f to $|\psi\rangle$ evaluates $f(x)$ for all x in parallel with

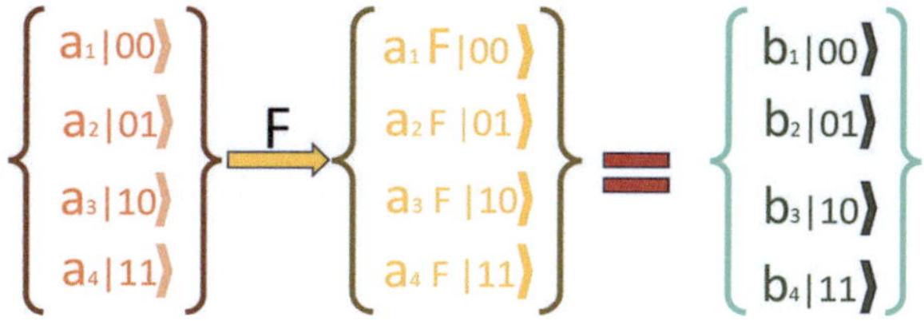

Fig. 3.5 Quantum parallelism

a single circuit. This massive parallelization can provide exponential speedups for specific problems and algorithms (Fig. 3.5).

However, directly observing the output would collapse the state. Special techniques like interference are needed to extract beneficial results. Quantum parallelism provides the foundation for quantum algorithms to outperform classical counterparts.

References

1. Nielsen, M., & Chuang, I. (2010). *Quantum computation and quantum information*. Cambridge University Press.
2. Born, M. (1962). Quantum mechanics of collision processes. *Science, 137*(3533), 683–684.
3. Pryde, G. (2005). Quantum interference. *Physics World, 18*(2), 33–37.
4. Horodecki, R., et al. (2009). Quantum entanglement. *Reviews of Modern Physics, 81*(2), 865–942.
5. von Neumann, J. (1955). *Mathematical foundations of quantum mechanics*. Princeton University Press.
6. Deutsch, D. (1985). Quantum theory, the Church–Turing principle and the universal quantum computer. In *Proceedings of the Royal Society of London. Series A: Mathematical, Physical, and Engineering Sciences, 400*, 1818.

Chapter 4
Quantum Circuits

Quantum circuits provide a model for describing and implementing quantum algorithms by applying operations to qubits arranged in wires. This chapter introduces the basic structure of quantum circuits, surveys standard quantum logic gates, walks through writing simple quantum code, and explains how simulator tools can be used to experiment with quantum circuits classically.

4.1 Basic Structure of Quantum Circuits

Quantum circuits offer a framework for representing sequences of operations on qubits analogous to logic circuits for classical computers [1]. Quantum circuits consist of qubits (wires) that information flows through, connected by quantum logic gates representing operations, assembled into circuits (algorithms). Some basic principles for quantum circuit diagrams include:

- Time flows left to right, with gates applied sequentially.
- Horizontal wires represent qubits. N qubits are stacked vertically.
- Gates describes unitary operations applied to target qubits.
- Measurement operations are depicted by meters at the circuit end.

For example, a simple 2-qubit quantum circuit (Fig. 4.1).

The notation $|\psi\rangle$ represents an arbitrary input quantum state fed into the qubits. U_1 and U_2 are unitary gates operating on each qubit. The final measurements collapse the state. Multiple gates and measurements can be included (Fig. 4.2).

M. Singh and B. S. Rawal, *Quantum Minds: Exploring Quantum Computing and AI Synergy*, https://doi.org/10.1007/978-981-95-2498-3_4

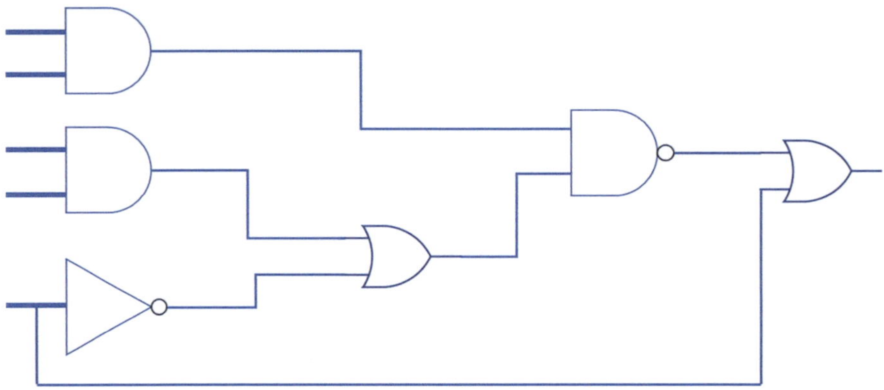

Fig. 4.1 Simple 2-qubit quantum circuit

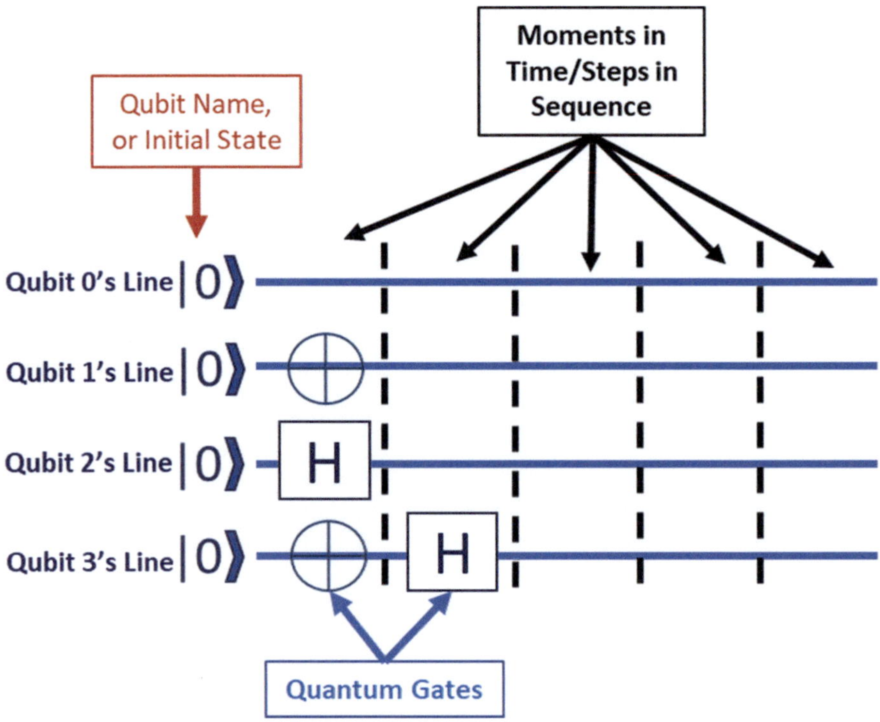

Fig. 4.2 Basic quantum circuit structure

4.2 Common Quantum Logic Gates

Quantum algorithms are constructed by composing a series of elementary quantum logic gates that perform basic unitary operations. Here, we introduce some of the most essential single- and multi-qubit quantum gates [2]:

Hadamard (*H*) Gate

Applies a Hadamard transform that puts a qubit into equal superposition of basis states (Fig. 4.3).

$$H|0\rangle = \left(1/\sqrt{2}\right)(|0\rangle + |1\rangle)$$

$$H|0\rangle = \left(1/\sqrt{2}\right)(|0\rangle - |1\rangle)$$

Pauli (*X*, *Y*, *Z*) Gates

The Pauli gates apply σ_x, σ_y, and σ_z operators that flip or phase shift qubit basis states (Fig. 4.4).

X NOT gate, flips $|0\rangle \leftrightarrow |1\rangle$.
Y bit and phase flip gate.
Z phase flip gate.

Phase (*S*, *T*) Gates

Apply phase shifts to quantum states:

S Phase gate, $|0\rangle \to |0\rangle, |1\rangle \to i|1\rangle$.
T $\frac{\alpha}{4}$ phase rotation gate.

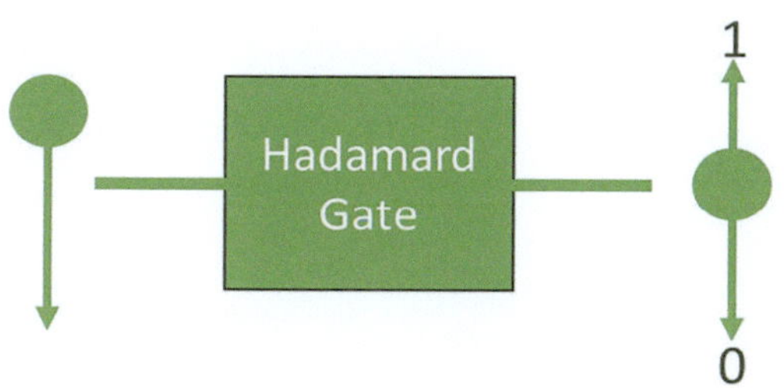

Fig. 4.3 Hadamard gate

Fig. 4.4 Pauli gate

Rotation (R_n) Gates

Rotate a qubit's state vector on the Bloch sphere by angle θ (Fig. 4.5).

$$R_n|\psi\rangle = \mathrm{e}^{\left(i\frac{\Theta}{2}\right)}|\psi\rangle$$

Controlled NOT (CNOT) Gate

Flips target qubit (control = 1) based on control qubit value: $|x\rangle|y\rangle \rightarrow |x\rangle|x \oplus y\rangle$ (Fig. 4.6).

Toffoli (CCNOT) Gate

Flips target qubit if both control qubits are set: $|x\rangle|y\rangle|z\rangle \rightarrow |x\rangle|y\rangle|z \oplus (x \cdot y)\rangle$ (Fig. 4.7).

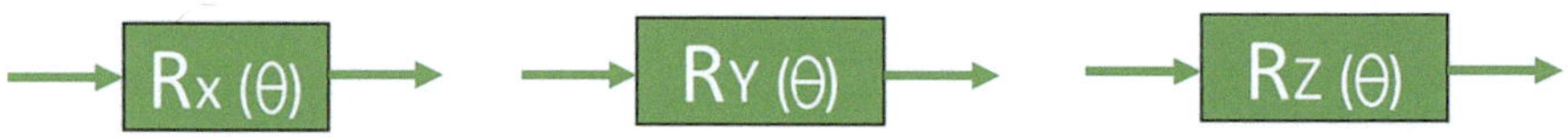

Fig. 4.5 Rotation (R_n) gates

Fig. 4.6 Controlled NOT (CNOT) gate

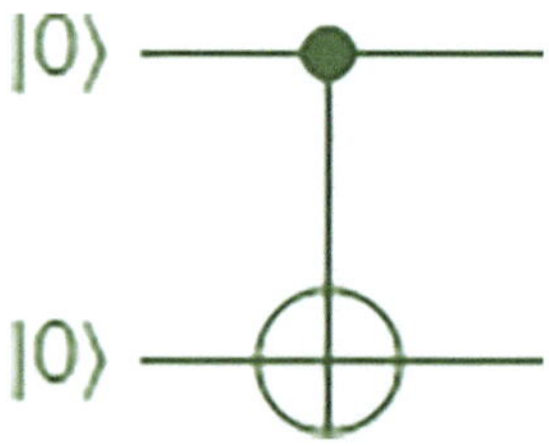

Fig. 4.7 Toffoli (CCNOT) gate

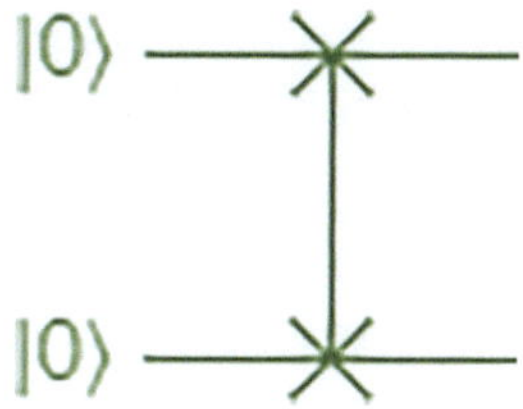

Fig. 4.8 Swap gate

Swap Gate

Swaps states between two qubits: $|x\rangle|y\rangle \rightarrow |y\rangle|x\rangle$ (Fig. 4.8).

When composed together, these gates form a universal set that can approximate any unitary operation to arbitrary accuracy. By layering these basic operations, more complex quantum programs can be constructed.

4.3 Writing Simple Quantum Circuits

We can now write simple quantum code with an understanding of basic gates. For example, here is how to construct a 1-qubit Grover diffusion operator circuit [3]:

- *H gate—Put qubit in equal superposition*
- *X gate—Invert amplitudes*
- *H gate—Rotate superposition*
- *Z gate—Rotate superposition further*
- *H gate—Complete rotation*

In circuit diagram form:

We can see the qubits initialized, operated on by a sequence of gates, then measured. Multiple qubits and gates build up complex quantum programs. Let's analyze some examples of pseudocode for generating quantum circuits in Python using shared libraries (Fig. 4.9).

Start by importing tools and initializing some qubits

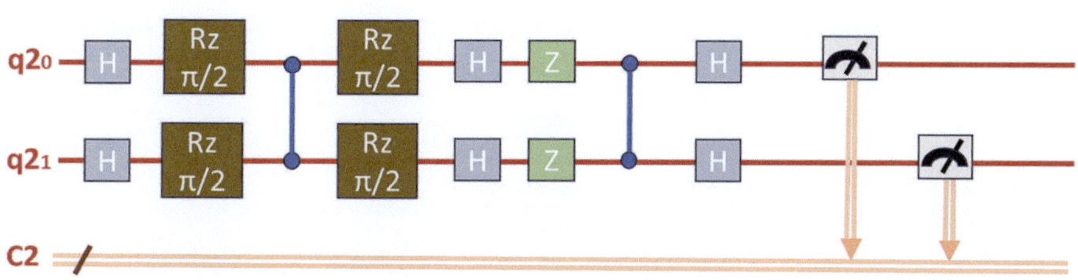

Fig. 4.9 Simple quantum circuit

```
from qiskit import QuantumCircuit, QuantumRegister
from qiskit.extensions import Initialize
qr = QuantumRegister(2, 'qr')
circuit = QuantumCircuit(qr)
```

Next, we apply some gates to our qubits

```
circuit.h(qr[0]) # Apply H gate to qubit 0
circuit. cx(qr[0], qr[1]) # CNOT with qubit 0 as control, one as target
circuit.barrier()
circuit.rz(pi/4, qr[1]) # Z rotation of qubit 1
circuit.measure_all()
```

Visualize the circuit

```
circuit.draw(output= 'mpl')
```

And run simulations using Qiskit's Aer simulator

```
from qiskit import BasicAer, execute
backend = BasicAer.get_backend('statevector_simulator')
job = execute(circuit, backend)
result = job.result()
print(result.get_statevector(circuit))
```

This workflow allows constructing, visualizing, and classically simulating quantum circuits to understand their behavior.

4.4 Quantum Circuit Simulators

Quantum computers are still quite limited in qubit number, connectivity, and coherence times. We can leverage quantum circuit simulators to experiment with more extensive quantum algorithms and circuits precluded by current hardware. These simulators mathematically simulate the expected results of quantum circuits on classical hardware [4].

Popular open-source quantum circuit simulators include:

- ***Qiskit Aer***—Simulators from IBM for executing circuits locally https://qiskit.org/ecosystem/aer/

- ***Cirq***—Google's Python framework with built-in state vector simulators https://quantumai.google/cirq

- ***Quantum++***—High performance C++ CPU/GPU-based simulator https://formulae.brew.sh/formula/quantum++
- ***Quirk***—Browser-based drag-and-drop quantum circuit simulator https://algassert.com/quirk
- ***Quantum Learning Machine***—Cloud-based circuit simulator focusing on education https://atos.net/en/solutions/quantum-learning-machine

These tools solve Schrödinger's equation to determine expected state vector outputs for quantum circuits without requiring actual qubits. Some benefits include:

- Test circuits are destined for real quantum hardware.
- Debug and verify the correctness of quantum programs.
- Model large circuits exceeding current qubit limits.
- Perform controlled experiments with noise models.
- Develop quantum algorithms before real devices mature.

While not a perfect substitute for physical qubits, simulators enable expedited development cycles and the building of quantum programming skills as the field continues maturing.

References

1. Nielsen, M., & Chuang, I. (2010). *Quantum computation and quantum information.* Cambridge University Press.
2. IBM Quantum Experience Documentation. https://quantum-computing.ibm.com/docs/
3. Grassl, M., et al. (2014). Efficient quantum circuits for one-way quantum computing. *Advances in Applied Mathematics, 60*, 1–36.
4. Guerreschi, G. G., & Matsuura, A. Y. (2020). QAOA for MaxCut requires hundreds of qubits for quantum speed-up. *Scientific Reports, 10*(1).

Chapter 5
Executing Circuits on Real Devices

5.1 Connecting to Quantum Hardware

While simulators provide a valuable development environment, ultimately, we want to run quantum circuits on actual quantum hardware to harness true quantum effects. Real quantum computers are accessible via cloud platforms from providers like IBM, Rigetti, Amazon, and Microsoft [1] (Fig. 5.1).

These cloud-based services provide authenticated remote access to prototype quantum processors comprising stable cold atoms, trapped ions, or superconducting qubits. Compilation tools map quantum code to the target device topology and set available gates. Experiments run as jobs on quantum hardware, with runtime priority and quotas.

For example, connecting to IBM's cloud quantum devices in Python

```
From qiskit import IBMQ
IBMQ.load_account() # Load IBMQ credentials
provider = IBMQ.get_provider()
backend = provider.get_backend('ibmq_16_melbourne') # Quantum processor
print(backend)
This prints:

    ibmq_16_melbourne (Melbourne)
    Backend properties:

Qubits: 15
```

We now have a backend connected for running quantum jobs. Experiments are orchestrated by (Fig. 5.2).

This workflow enables leveraging existing robust quantum hardware over the cloud.

M. Singh and B. S. Rawal, *Quantum Minds: Exploring Quantum Computing and AI Synergy*, https://doi.org/10.1007/978-981-95-2498-3_5

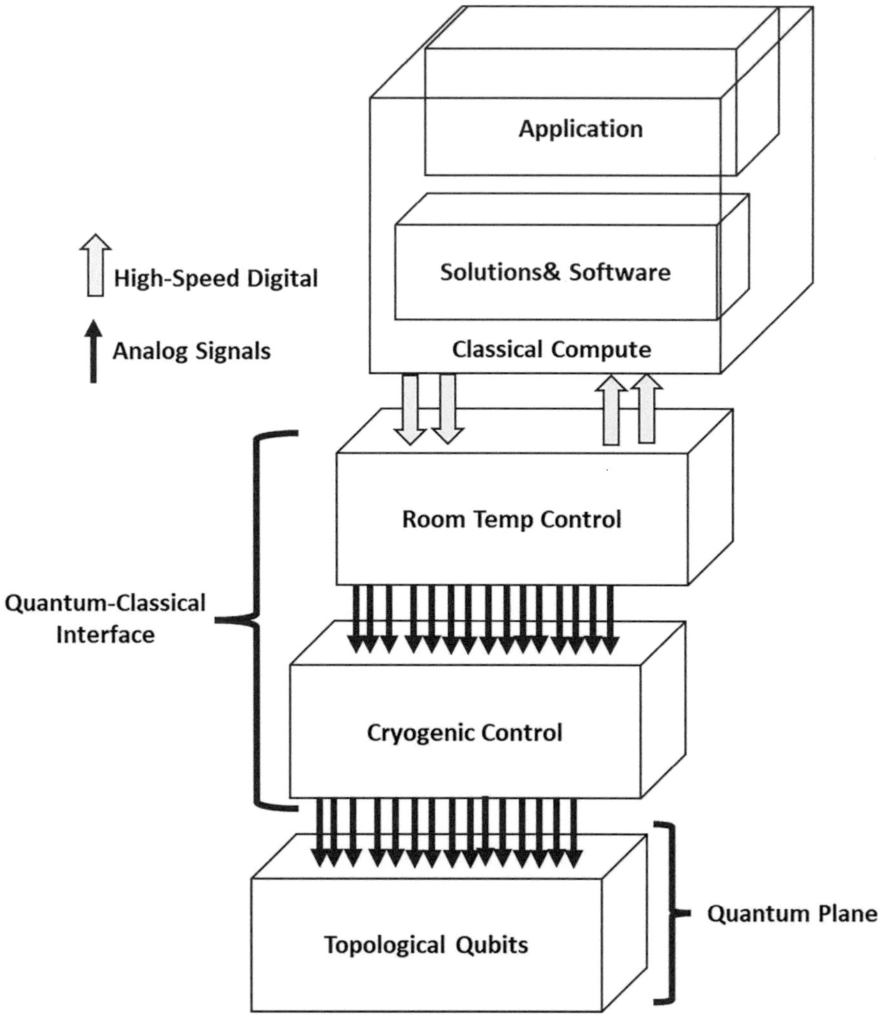

Fig. 5.1 Connecting to quantum hardware

5.2 Running Experiments on IBM Quantum Experience

The IBM Quantum Experience provides free public access to prototype IBM quantum systems for education and research [2]. Users can experiment with real quantum hardware from anywhere with an internet connection.

The steps to run circuits on IBM quantum devices are:

1. Sign up for an IBM Quantum Experience account.
2. Use the Composer tool to construct quantum circuits.
3. Select a backend simulator or quantum chip.

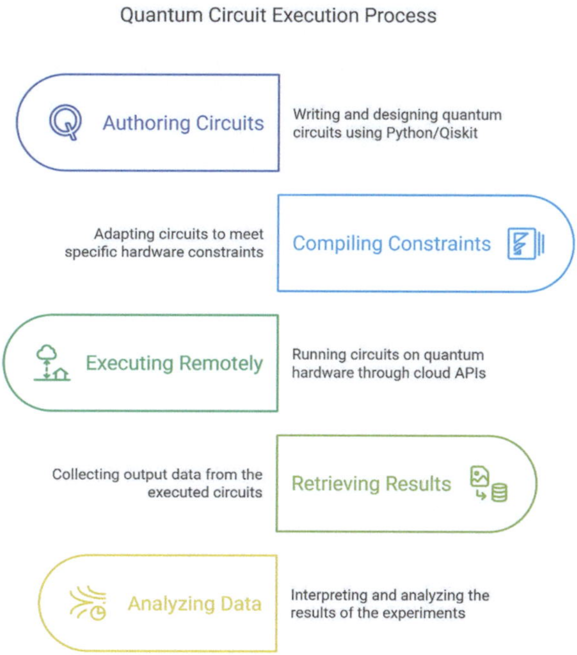

Fig. 5.2 Quantum circuit execution process

4. Compile and optimize the circuit for the target.
5. Submit a job to run an experiment.
6. Retrieve and display results when complete (Fig. 5.3).

For example, this Python code runs a circuit on the ibmq_qasm_simulator

```
From qiskit, import Aer, transpile, and assemble
from qiskit. providers.ibmq import least_busy
backend = Aer.get_backend('ibmq_qasm_simulator')
circuit = QuantumCircuit(qr, cr) # Created earlier
transpiled_circuit = transpile(circuit, backend)
qobj = assemble(transpiled_circuit)
job = backend.run(qobj)
job_monitor(job)
results = job.result()
print(results.get_counts(transpiled_circuit))
```

We retrieve job results and can analyze the output. This demonstrates running quantum code on real hardware through IBM Quantum Experience.

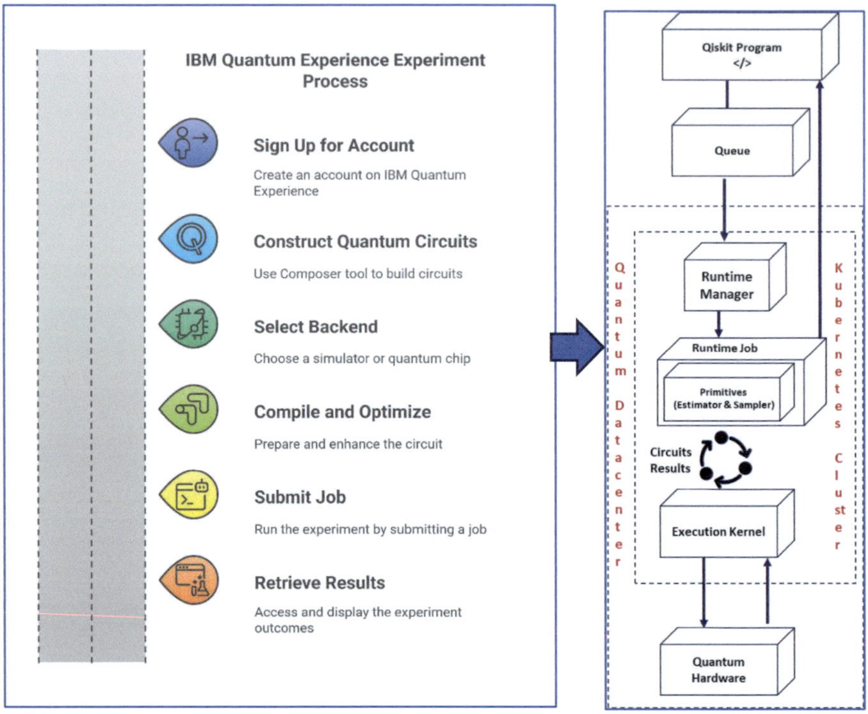

Fig. 5.3 Running experiments on IBM quantum experience

5.3 Running Experiments on Rigetti Quantum Cloud Services

Rigetti Computing offers access to its quantum processors, including its Aspen series superconducting qubit devices, through Rigetti Quantum Cloud Services [3].

Critical steps to running circuits on Rigetti devices include:

- Create an account on Rigetti Forest and obtain API keys.
- Install pyQuil and the Forest SDK locally.
- Connect to a Rigetti backend target like Aspen-9.
- Author quantum circuits in PyQuil or Quil.
- Compile quil to native backend gates.
- Execute job on quantum processor.
- Retrieve results and quil execution details when complete (Fig. 5.4).

For example, this PyQuil code submits a Bell state experiment

```
from pyquil import Program, get_qc
from pyquil.gates import H, CNOT
```

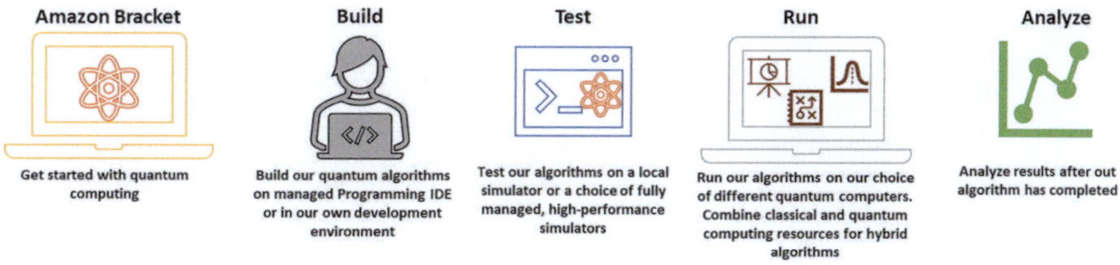

Fig. 5.4 Running experiments on rigetti quantum cloud services

```
from pyquil.API import WavefunctionSimulator, QVMConnection
shots = 1000
qvm = QVMConnection()
backend = qvm.get_backend('Aspen-9')
p = Program(H(0), CNOT(0, 1))
executable = p.compile(backend)
job = qvm.run(executable, shots=shots)
result = job.result()
print(result)
```

We can run circuits on Rigetti's quantum chips and simulators through Forest's APIs.

5.4 Troubleshooting Errors

Various types of errors can arise when executing quantum programs, affecting the fidelity of results [4]. Understanding and debugging these errors is an important skill (Figs. 5.5 and 5.6).

Common quantum hardware errors include:

- ***Qubit decoherence***: qubits lose state purity over time due to spontaneous emission and environmental noise.
- ***Gate errors and*** imprecise gates interact with qubits in unintended ways.
- ***Measurement errors***, incorrect qubit measurement outcomes.
- ***Qubit connectivity***, limited qubit interactions imposed by hardware graphs.

Debugging techniques include:

- ***Noise modeling***, simulating theoretical noise profiles to predict likely errors.
- ***Tomography***, reconstructing quantum states to assess deviations.
- ***Adding redundancies and*** repetition codes to detect errors through voting.
- ***Validating subsystems and*** test components like gates independently.
- ***Monitoring operations and*** collecting performance metrics across the software and hardware ***stack***.

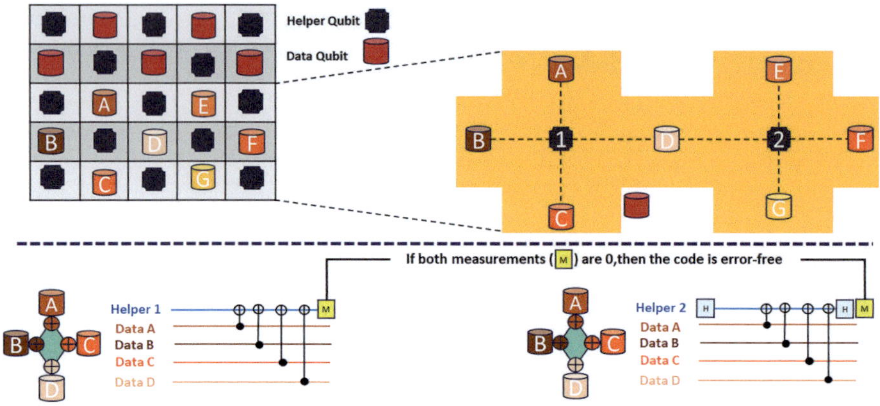

Fig. 5.5 Troubleshooting errors

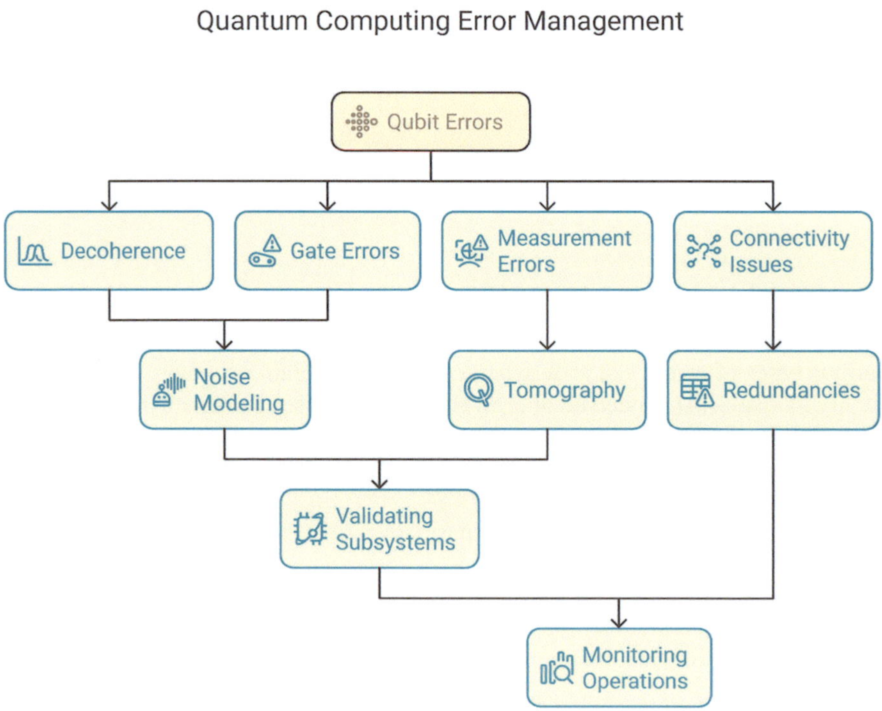

Fig. 5.6 Troubleshooting errors

With careful characterization and debugging, sources of errors can be identified and reduced to improve results. This is an active area of ongoing research as quantum hardware scales up.

References

1. Preskill, J. (2018). Quantum computing in the NISQ era and beyond. *Quantum, 2*, 79.
2. Cross, A., et al. (2019). Validating quantum computers using randomized model circuits. *Physical Review A, 100*(3).
3. Karalekas, P., et al. (2020). A quantum-classical cloud platform optimized for variational hybrid algorithms. *Quantum Science and Technology 5*(2).
4. Fu, X., et al. (2007). Experimental demonstration of a compiled version of shor's quantum factoring algorithm using photonic qubits. *Physical Review Letters, 99*(25).

Chapter 6
Quantum Algorithms

Quantum algorithms leverage the unique capabilities of quantum systems to solve some computational issues exponentially faster than the best-known classical approaches. This chapter explains several seminal quantum algorithms, including the Deutsch-Jozsa, Grover's search, Shor's factoring, and variational quantum eigensolver algorithms (Fig. 6.1).

6.1 Deutsch-Jozsa Algorithm

The Deutsch-Jozsa algorithm exemplifies how quantum parallelism provides an advantage over classical algorithms, even without entanglement [1]. It determines whether a function f mapping $f\{0, 1\}n \rightarrow \{0, 1\}$ is constant (all equal outputs) or balanced (half 0 and half one outputs) with certainty in just one evaluation (Fig. 6.2).

Classically, this requires $2^{n-1} + 1$ evaluations in the worst case to distinguish between constant and balanced conclusively. However, the quantum algorithm succeeds with only one function query using superposition.

Quantum Circuit

1. Initialize n qubits to $|0\rangle$ and one output qubit to $|1\rangle$.
2. Apply Hadamard gates H to each input qubit, putting them in an equal superposition of 0 and 1.
3. Query Oracle function f, which marks the output qubit: $|x\rangle|y\rangle \rightarrow |x\rangle|y \oplus f(x)\rangle$.
4. Apply H gates again to the input register.
5. Measure all qubits.

If the output qubit is $0, f$ is constant. If $1, f$ is balanced. In either case, the input register will be a uniform superposition as shown in Fig. 6.3.

M. Singh and B. S. Rawal, *Quantum Minds: Exploring Quantum Computing and AI Synergy*, https://doi.org/10.1007/978-981-95-2498-3_6

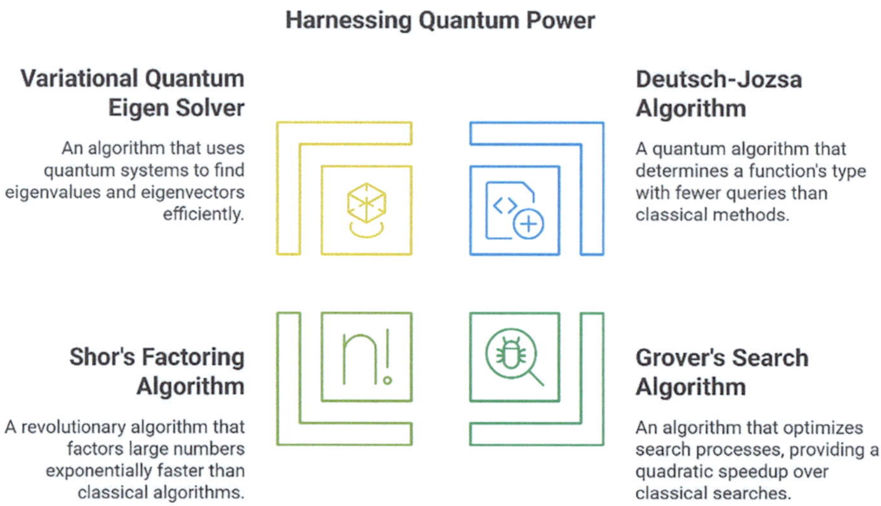

Fig. 6.1 Quantum computing algorithms

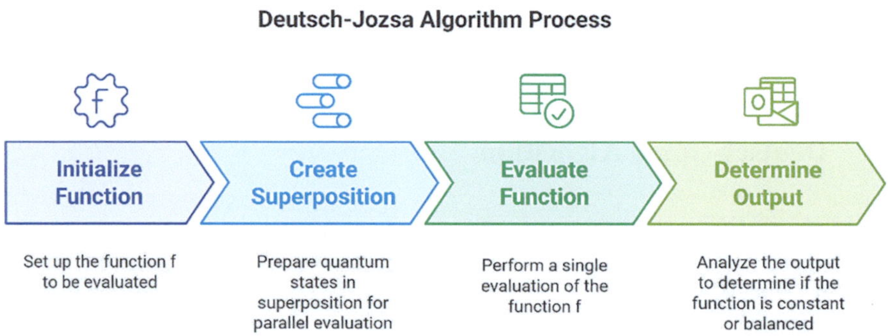

Fig. 6.2 Deutsch-Jozsa algorithm

Fig. 6.3 Example of Deutsch-Jozsa algorithm

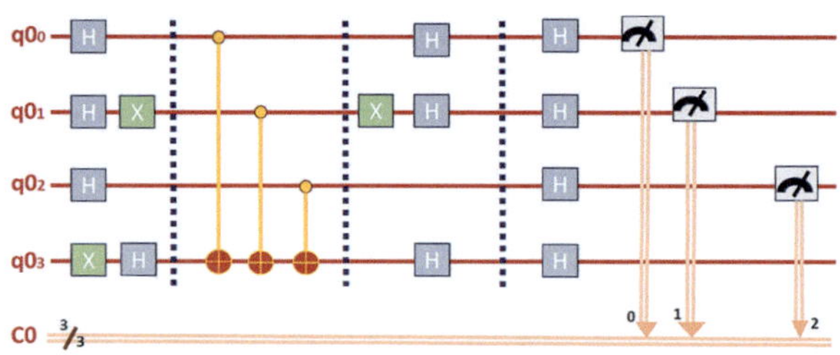

Pseudocode

Set 'n' input qubits to state |0⟩
Set output qubit to state |1⟩
Apply Hadamard gate to each input qubit
 # Query Oracle function f
for x in a superposition of all n-bit inputs:
  if f(x) = 1:
    Flip output qubit
Apply Hadamard gate to each input qubit
  Measure all qubits
  If the output qubit is 0:
    Function is constant
  Else:
    Function is balanced

By leveraging a quantum superposition to evaluate $f(x)$ for all values of x in parallel, the Deutsch-Jozsa algorithm deterministically solves the problem with one function evaluation instead of up to $2n$ classically.

6.2 Grover's Algorithm

Grover's algorithm provides a quadratic speedup for searching unsorted databases over classical approaches [2]. Given an unsorted database of N items, it can find the index of a target item using only $O\sqrt{N}$ queries compared to $O(N)$ classically as represents in Fig. 6.4.

The algorithm works by utilizing an oracle function, phase shifts, and inversions to amplify the amplitude of the target state selectively.

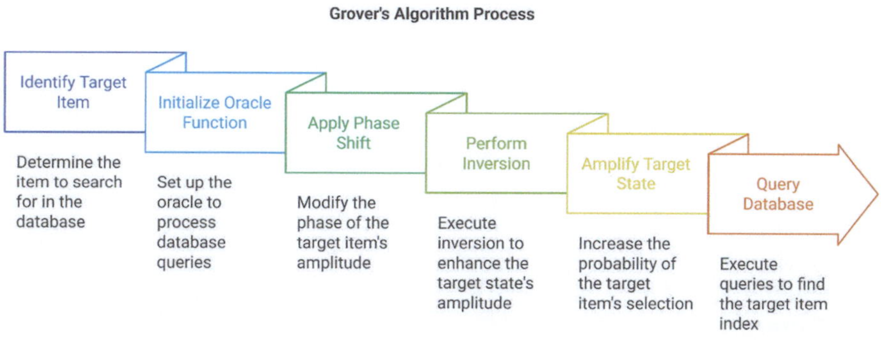

Fig. 6.4 Grover Algorithm

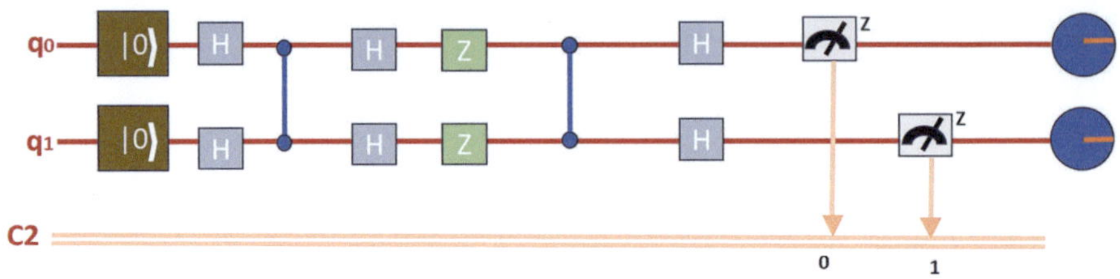

Fig. 6.5 Example of Grover's algorithm

Steps:

a. Initialize register to uniform superposition.
b. Repeat $O\sqrt{N}$ times:

   1. Query Oracle to mark the target state.
   2. Invert phases.
   3. Reflect over mean.

Measure register

This iteratively increases the probability that measuring the register reveals the target index Fig. 6.5 shown an example of Grover algorithm.

Pseudocode

```
import math
N = len(database)
target_index = 42 # Item to find
qubits = ceil(log2(N))
# Initialize uniform superposition
for qubit in range(qubits):
   ApplyH(qubit)
iterations = math.ceil(math.sqrt(N))
for i in range(iterations):
   # Oracle marks the target state
  if register == target_index:
   ApplyConditionalPhaseShift()
  # Inversion about mean
  ApplyZ()
  ApplyH()
  ApplyZ()
  ApplyH()
# Measure register
Measure ()
print ("Target index:", result)
```

Grover's algorithm achieves a quadratic speedup over classical counterparts by exploiting quantum parallelism and interference.

6.3 Shor's Algorithm

Shor's algorithm enables integer factorization in polynomial time using quantum parallelism, providing an exponential speedup over the best-known classical factoring algorithms [3]. The ability to quickly factor large numbers could break modern public key cryptography like RSA; it is represented in Fig. 6.6.

Steps are:

1. *Pick a random number, $a < N$, where N is the number to be factored.*
2. *Compute order r of a modulo N efficiently using a period-finding subroutine.*
3. *Solve $a^{r/2} = -1$ (mod N) using gcd to find non-trivial factors of N*

The key innovation is the period-finding subroutine. Classically, this takes $O(2n)$ but only the quantum of $O(n^3)$. It consists of:

1. Put registers in superposition representing all x and y.
2. Compute modular exponentiation function $f(x) = a^x \bmod N$.
3. Apply quantum Fourier transform to the x register.
4. Measure registers to observe periodicity.

This provides the order r needed to factor N. Shor's algorithm can efficiently factor large numbers intractably classically by exploiting superposition and Fourier transforms. The implications for cryptography are concerning shown in Fig. 6.7.

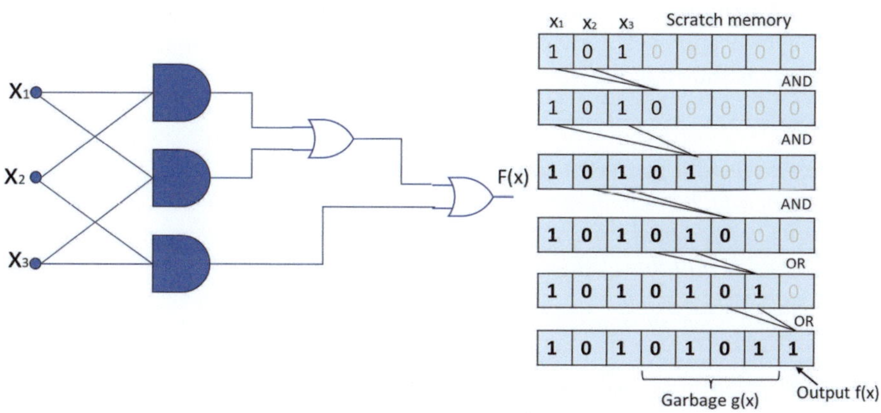

Fig. 6.6 Example of Shor's algorithm

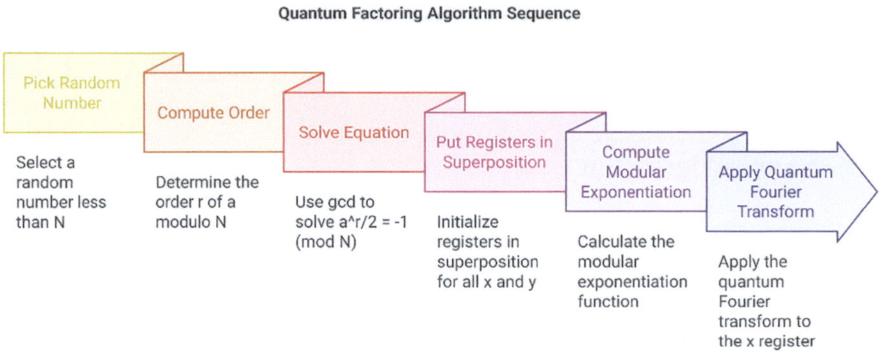

Fig. 6.7 Quantum factoring algorithm

6.4 Variational Quantum Eigensolver (VQE)

The variational quantum eigensolver (VQE) is a leading hybrid classical-quantum algorithm used in quantum chemistry for finding molecular ground state energies [4]. It works by:

1. Encoding molecular structure into parametrized qubit Hamiltonian
2. Initializing trial qubit state
3. Measuring energy of state
4. Using classical optimizer to vary parameters, minimizing energy
5. Returning optimal minimum energy state.

While the classical optimizer runs on a regular computer, preparing quantum states, computing energies, and measuring can be performed on noisy quantum hardware. Variational Quantum Eigensolver process represents in Fig. 6.8.

This hybrid approach provides an approximate solution faster than classical approaches, as shown in Fig. 6.9.

Pseudocode

```
# Map molecular system to qubit Hamiltonian
H = generate_hamiltonian(molecule)
# Parameterized trial state
trial_state = initialize_vqe_state(qubits, params)
best_parameters = []
best_energy = float('info')
for epoch in range(max_epochs):
  current_energy = evaluate_energy(trial_state)
  if current_energy < best_energy:
    best_energy = current_energy
    best_parameters = params
  # Classical optimization update
```

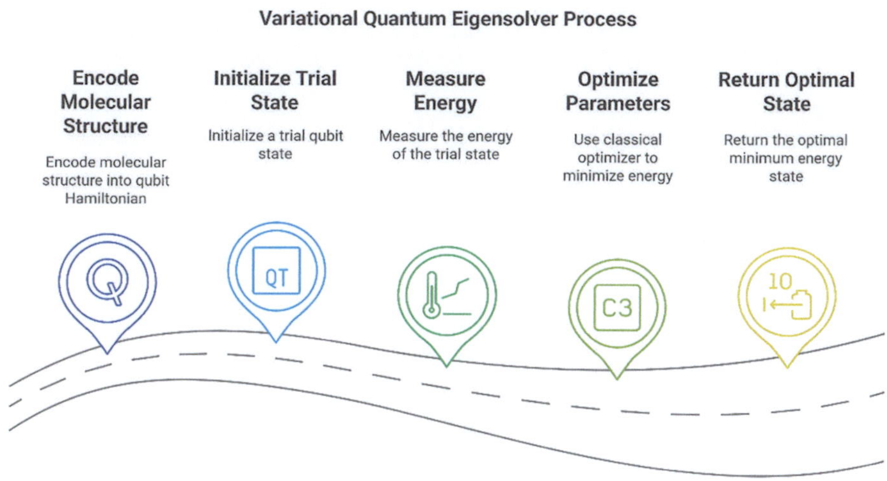

Fig. 6.8 Variational quantum eigensolver (VQE process)

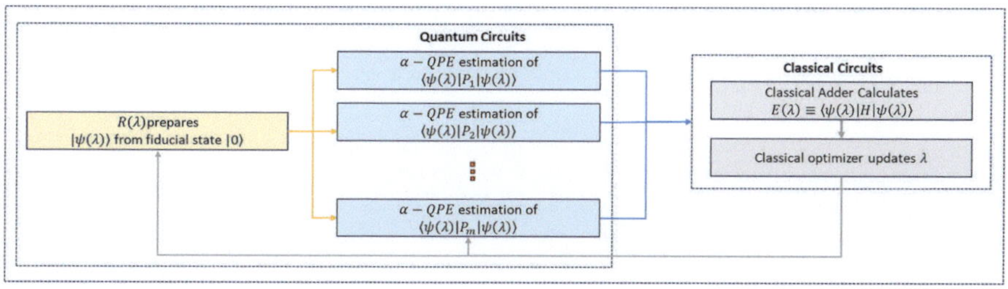

Fig. 6.9 Variation quantum eigensolver (VQE) example

params = optimizer.update(current_energy, params)
 # Reinitialize trial state with new parameters
trial_state = initialize_vqe_state(qubits, params)
return best_parameters, best_energy

While not achieving an exponential speedup, VQE uses quantum resources available today in a hybrid algorithmic approach. Quantum advantages will likely first emerge through such applications.

References

1. Deutsch, D., & Jozsa, R. (1992). Rapid solution of problems by quantum computation. *Proceedings of the Royal Society of London. Series A: Mathematical and Physical Sciences, 439*(1907), 553–558.

2. Grover, L. K. (1996). A fast quantum mechanical algorithm for database search. In *Proceedings of the twenty-eighth annual ACM symposium on theory of computing.*
3. Shor, P. W. (1997). Polynomial-time algorithms for prime factorization and discrete logarithms on a quantum computer. *SIAM Journal on Computing, 26*(5), 1484–1509.
4. Peruzzo, A., et al. (2014). A variational eigenvalue solver on a photonic quantum processor. *Nature Communications, 5*(1).

Chapter 7
Applications for Quantum Computing

While large, error-corrected quantum computers remain years away, today's noisy, intermediate-scale quantum (NISQ) devices drive innovation across many application domains. This chapter explores how quantum computing impacts quantum chemistry, optimization, finance, machine learning, and secure communications; its shown description in Fig. 7.1.

7.1 Quantum Chemistry Simulation

Quantum simulation applies to quantum computers to emulate the properties of other quantum systems. This is especially promising for studying chemical reactions and properties [1]. Electron behavior and reactions are driven by complex quantum interactions that stretch computational limits. Represents in Fig. 7.2.

Quantum chemistry simulation works by mapping properties like atomic structure and bonding into qubits and quantum gates to replicate a molecule's quantum Hamiltonian dynamics. Algorithms like the variational quantum eigensolver, phase estimation, and Trotterization are employed.

For example, simulating hydrogen

Map electron configuration to qubits $|01\rangle$.

- Approximate kinetic and potential energy terms with gates.
- Initialize superposition state.
- Measure the expectation value of Hamiltonian (Fig. 7.3).

This reveals the molecule's energy spectra and eigenstates. More complex molecules require more qubits and nuanced mapping.

Quantum computers hold great promise over classical approximation techniques for gaining insights into chemical reactions and pharmaceutical discovery by actually

M. Singh and B. S. Rawal, *Quantum Minds: Exploring Quantum Computing and AI Synergy*, https://doi.org/10.1007/978-981-95-2498-3_7

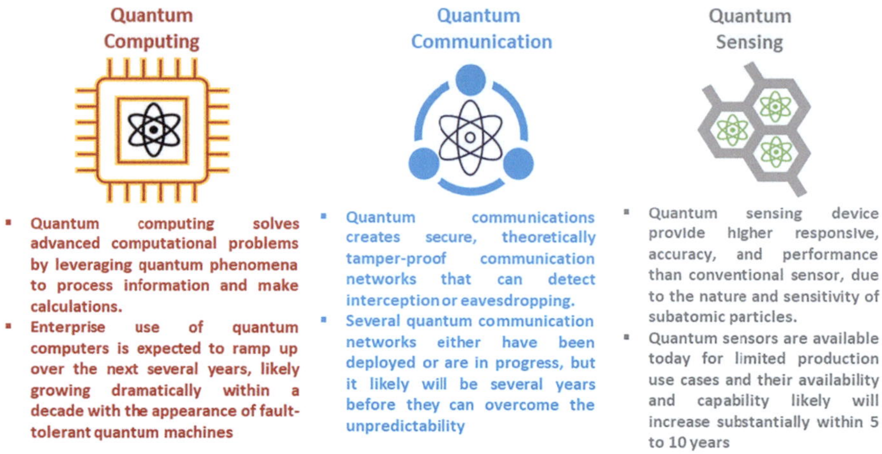

Fig. 7.1 Key quantum use cases

quantum mechanically representing quantum properties natively. Practical small-scale simulations have already been demonstrated on existing quantum hardware [2].

7.2 Optimization Using Quantum Annealing

Given a cost function, combinatorial optimization seeks optimal configurations, from scheduling to network routing. Quantum annealing utilizes quantum fluctuations to traverse search spaces for low-energy minima [3].

It first represents the optimization problem with a Hamiltonian whose ground state encodes the solution. The system is initialized in an equal superposition of all states. Then, quantum tunneling and adiabatic transitions are leveraged to evolve the state slowly toward lower energy configurations as represents in Fig. 7.4.

At the end of the anneal, measuring the qubits reveals bit values representing (hopefully) the globally optimal solution. D-Wave quantum annealing systems demonstrate this on real hardware containing over 5000 qubits.

7.3 Financial Analysis of Quantum Algorithms

Quantum computing may impact finance by speeding up certain computations, such as risk analysis, derivatives pricing, and process optimizations [4]. While awaiting maturity, today's NISQ devices demonstrate potential value in finance shown in Figs. 7.5 and 7.6.

Fig. 7.2 Quantum chemistry simulation

Applications include

Portfolio Optimization: Quantum annealing finds asset allocations that maximize return at a given risk tolerance. Solving Markowitz models leverages quantum optimization.

- *Pricing*—Quantum Monte Carlo methods for risk analysis and pricing exotic derivatives are faster than classical sampling.
- *Fraud Detection*—Quantum machine learning models flag credit card fraud faster with lower false positives.

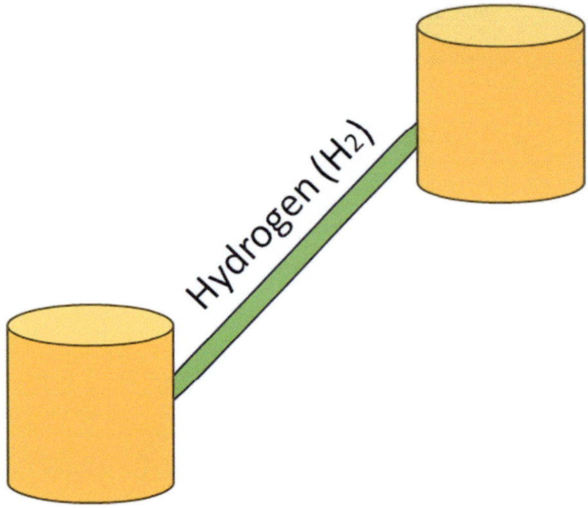

Fig. 7.3 Quantum chemistry simulation

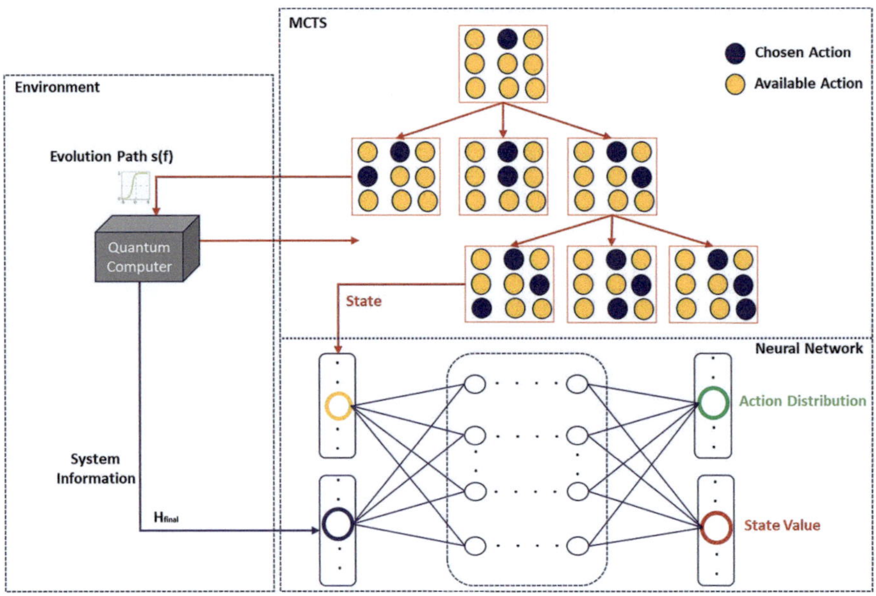

Fig. 7.4 Optimization using quantum annealing

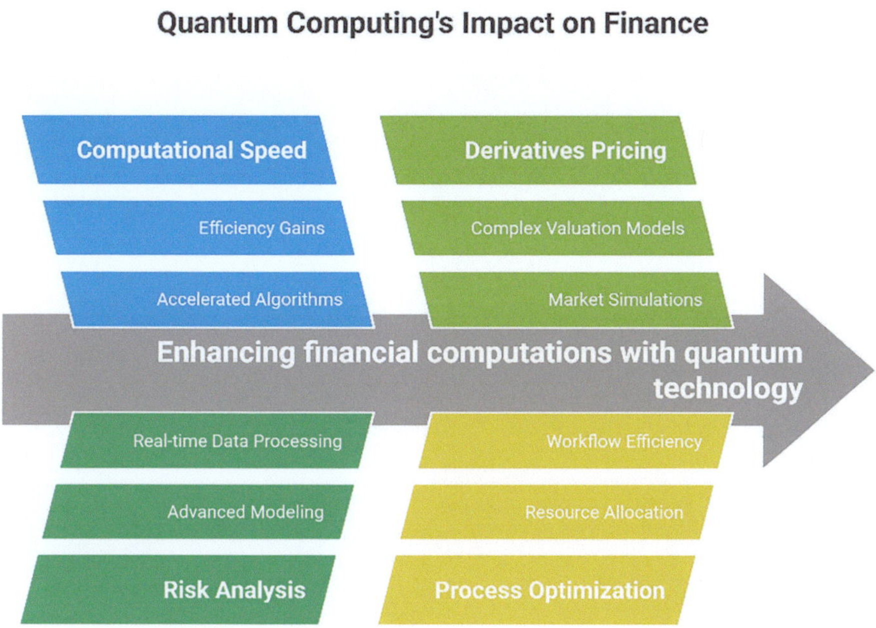

Fig. 7.5 Financial analysis of quantum computing

Fig. 7.6 Financial analysis with quantum algorithms

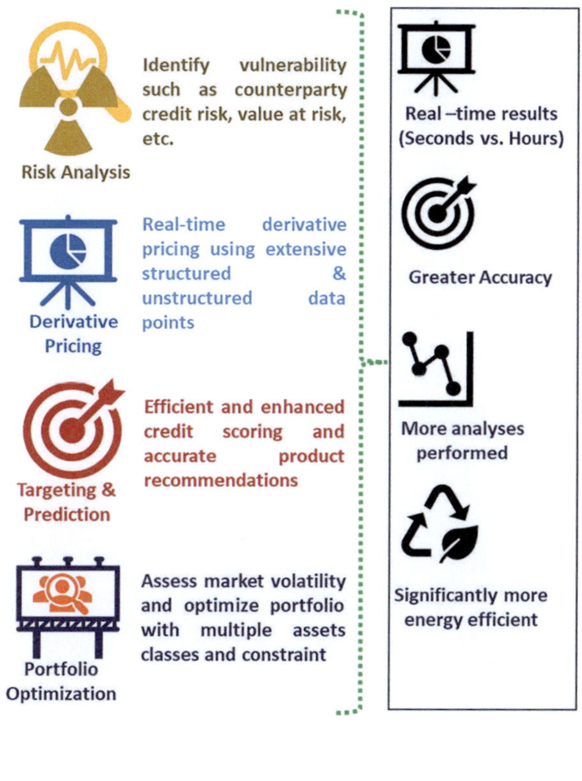

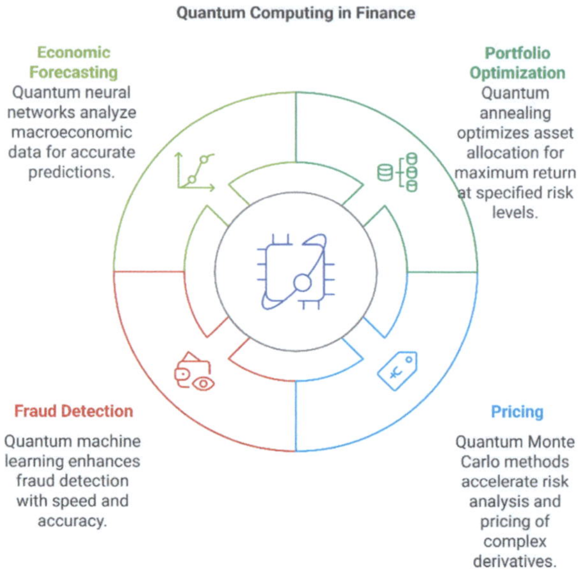

Fig. 7.7 Quantum computing in finance

- *Economic Forecasting*—Quantum neural networks can identify complex macroeconomic and trading signals.

Financial firms like JPMorgan and Goldman Sachs already experimented with near-term quantum potential [5]. As algorithms and hardware improve, quantum advantages for finance should grow presents in Fig. 7.7.

7.4 Quantum Machine Learning

Quantum machine learning explores using quantum computers to accelerate and enhance machine learning tasks like classification, regression, and clustering [6]. Rather than faster training, quantum aims for faster prediction and reasoning after training on classical hardware explanation presents in Fig. 7.8.

Two approaches include:

- **Quantum Neural Networks**
  Map neural network structures onto quantum circuits by encoding weights, activations, and training data into qubits and gates. Recurrent, convolutional, and feed-forward architectures are possible. Parameterized circuits learn complex patterns are represented in Fig. 7.9.

- **Quantum Kernel Methods**

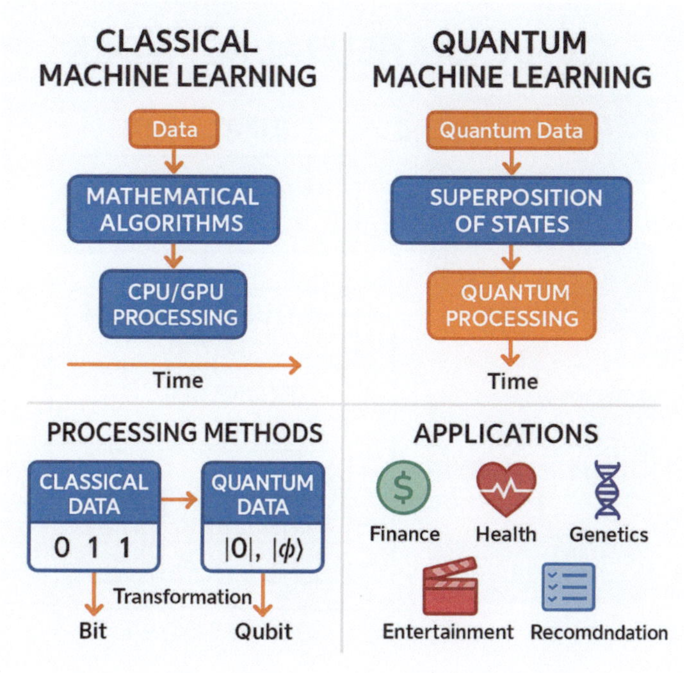

Fig. 7.8 Hybrid quantum machine learning

Use quantum circuits as kernel functions to map inputs into higher dimensional feature spaces for enhanced separability. Superposition enables the evaluation of inner products between all pairs simultaneously.

Challenges include noise and decoherence in current hardware and limited circuit size. However, improvements in quantum machine learning models matched with maturing physics hold promise for the future as shown in Fig. 7.10.

7.5 Secure Communications with Quantum Cryptography

Quantum physics enables two provably secure communication paradigms challenging to crack without detecting eavesdropping [7] as shown in Fig. 7.11.

Quantum Key Distribution (QKD)—Use photon polarization to share random secret keys between remote parties to enable one-time pad encryption. Any measurement collapses the state, revealing the hack as Fig. 7.12 represents.

For example, QKDs (Fig. 7.13)

1. Alice and Bob get entangled photon pairs.

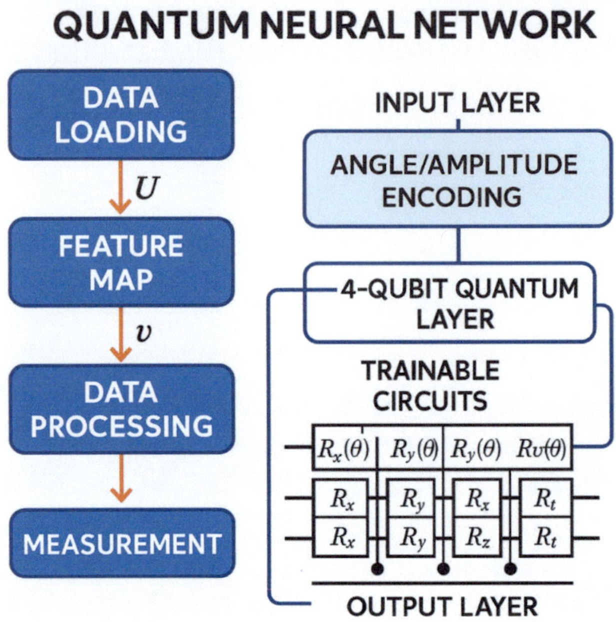

Fig. 7.9 Quantum neural networks

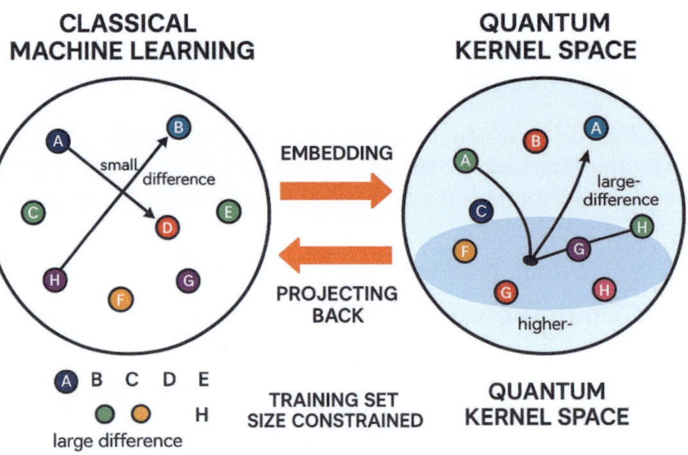

Fig. 7.10 Quantum kernel methods

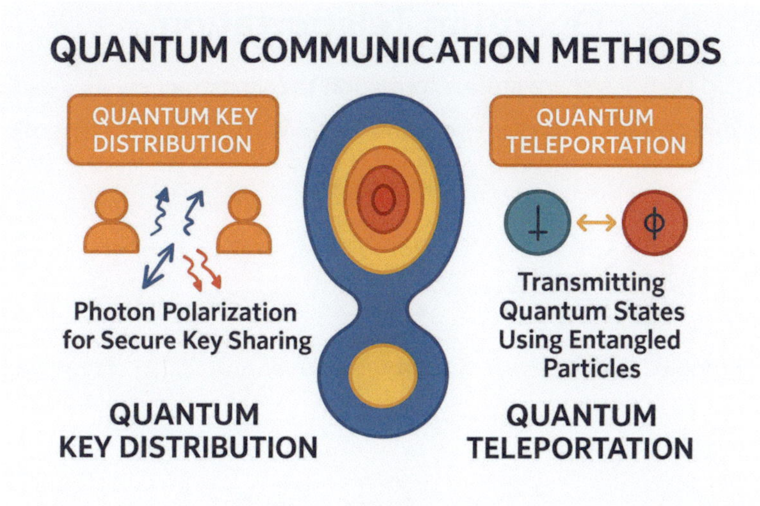

Fig. 7.11 Quantum communication methods

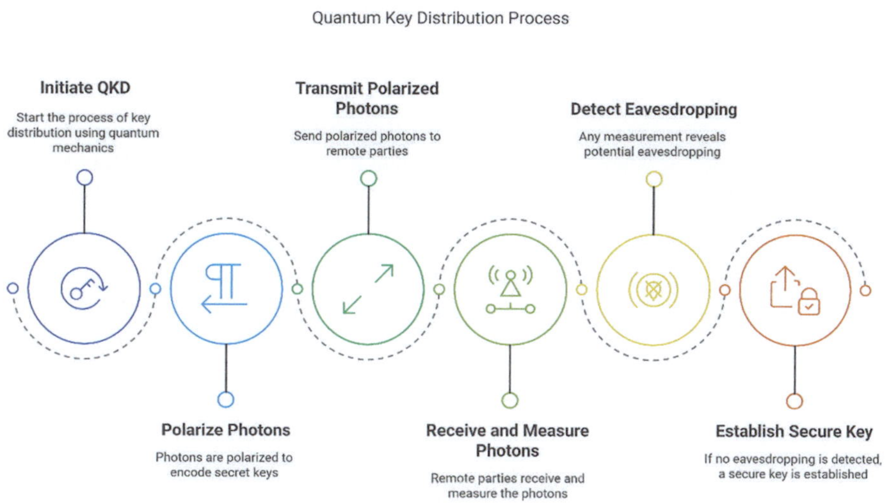

Fig. 7.12 Quantum key distribution process

2. Alice encodes her key into photons and sends it to Bob.
3. Bob measures photons, collapsing state.
4. Repeated key bits allow extracting shared keys over a public channel through reconciliation.
5. Any man-in-the-middle attack is detectable.

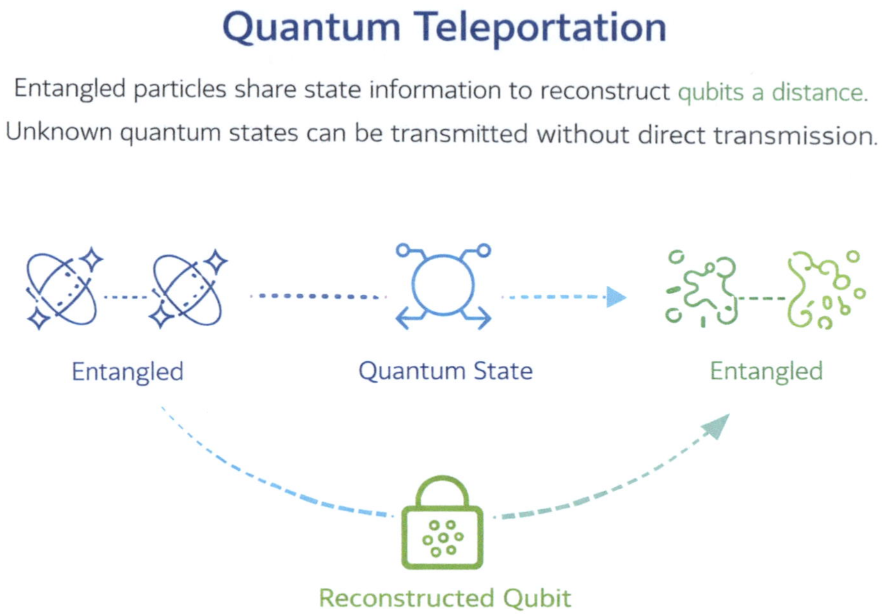

Fig. 7.13 Secure communications with quantum cryptography

QKD is already commercially deployed for high-security communications. As quantum networks advance, more advanced quantum cryptography protocols will emerge.

Quantum Teleportation—Entangled particles share state information to reconstruct qubits at a distance. Unknown quantum states can be transmitted without direct transmission (Fig. 7.14).

Fig. 7.14 Quantum teleportation

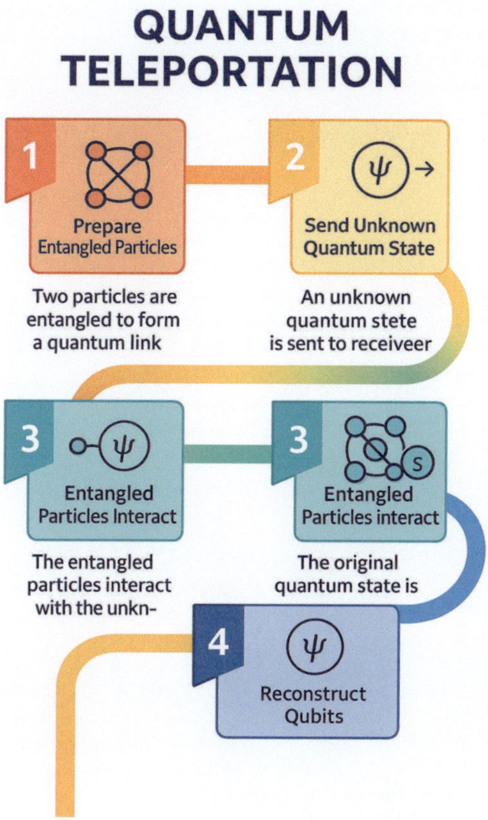

References

1. Cao, Y., et al. (2021). Quantum chemistry in the age of quantum computing. *Chemical Reviews, 121*(19), 10403–10454.
2. Kandala, A., et al. (2017). Hardware-efficient variational quantum eigensolver for small molecules and quantum magnets. *Nature, 549*(7671), 242–246.
3. Dash, A., et al. (2019). Exact combinatorial optimization by quantum annealing. arXiv preprint arXiv:1911.04058.
4. Orús, R., et al. (2020). Quantum computing for finance: state of the art and future prospects. *Reviews in Physics, 5*, 100028.
5. Raidl, G. R., & Mazzariol, F. (2021). Quantum computing and applications in finance. In *Proceedings of ECMFA*.
6. Schuld, M., et al. (2014). An introduction to quantum machine learning. *Contemporary Physics, 56*(2), 172–185.
7. Pirandola, S., et al. (2020). Advances in quantum cryptography. *Advances in Optics and Photonics, 12*(4), 1012–1236.

Part II
Quantum AI Revolution

Rapid progress in quantum computing and artificial intelligence represents two of our most disruptive technology trends. As these fields continue maturing, we witness their convergence into an exciting new paradigm of quantum-enhanced AI. This part provides a primer on quantum machine learning, which aims to augment classical AI algorithms with quantum processing capabilities.

Quantum machine learning explores how quantum computing principles can speed up traditionally slow machine learning tasks like training neural networks or traversing decision trees. Early theoretical results and proof-of-concept experiments demonstrate that quantum versions of supervised, unsupervised, and reinforcement learning protocols can achieve exponential speedups compared to classical implementations. Realizing these promising speedups will require surmounting formidable challenges in translating quantum ML theories into practical systems. Nonetheless, this emerging field provides a glimpse into the future quantum AI revolution.

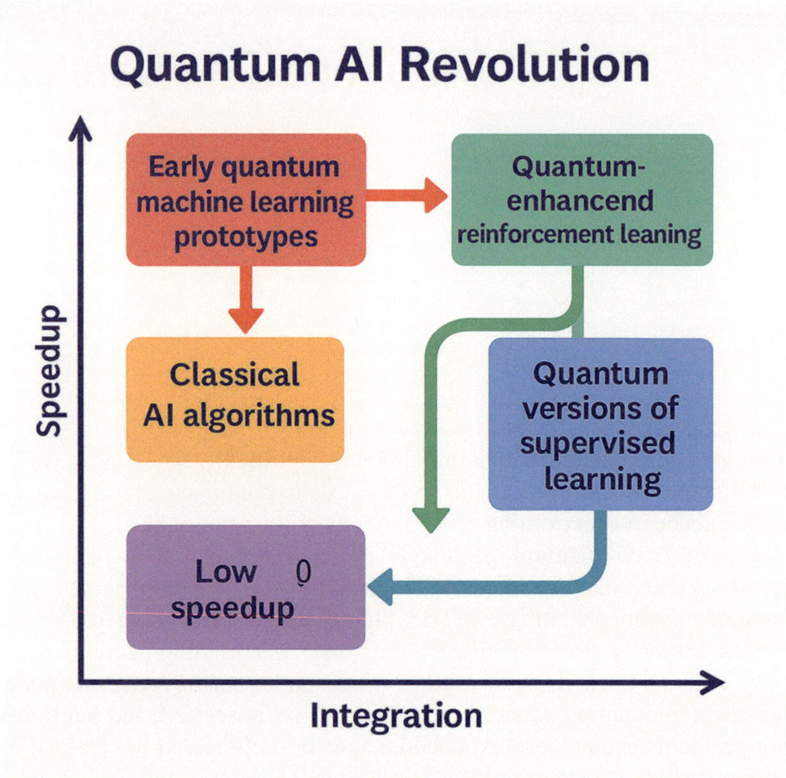

Fig. 2 Quantum artificial intelligence revolution

Chapter 8
Quantum Computing with Artificial Intelligence: Introduction

The past decade has witnessed rapid advances in two of the most transformative technologies of our time—quantum computing and artificial intelligence (AI) [1]. Though initially progressing along parallel tracks, these fields are now converging in exciting ways to create a new paradigm of computing that promises to reshape everything from drug discovery to financial modeling [2]. This chapter overviews this emerging field of quantum AI and discusses its potential relevance for industry management professionals.

Quantum computing harnesses the strange properties of quantum physics to perform calculations beyond the reach of even the most powerful supercomputers today. Rather than storing information as binary 0 s and 1 s like classical bits, quantum computers use quantum bits or qubits, which can exist in a superposition of 0 and 1 [3]. This allows qubits to perform multiple calculations simultaneously, enabling quantum computers to solve problems that would take conventional computers longer than the universe's age as shown in Fig. 8.1.

8.1 Understanding the Convergence of Quantum Computing and Artificial Intelligence

In recent years, quantum computing has graduated from a theoretical field to a rapidly maturing technology. Technology giants like IBM, Google, and Microsoft and startups like Rigetti and IonQ are developing practical quantum systems with 10–100 qubits [4]. At this scale, quantum computers can perform practical applications in chemistry, optimization, and machine learning. As quantum systems scale up to thousands or even millions of qubits in the coming decade, they are expected to deliver exponential leaps in processing power as shown in Fig. 8.2.

Parallel to these advances, AI has emerged as a revolutionary new approach to software development. AI systems utilize machine learning and neural networks to

M. Singh and B. S. Rawal, *Quantum Minds: Exploring Quantum Computing and AI Synergy*, https://doi.org/10.1007/978-981-95-2498-3_8

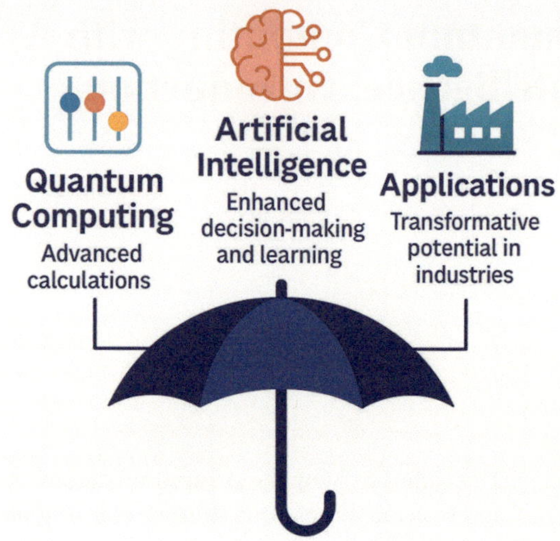

Fig. 8.1 Convergence of quantum AI computing

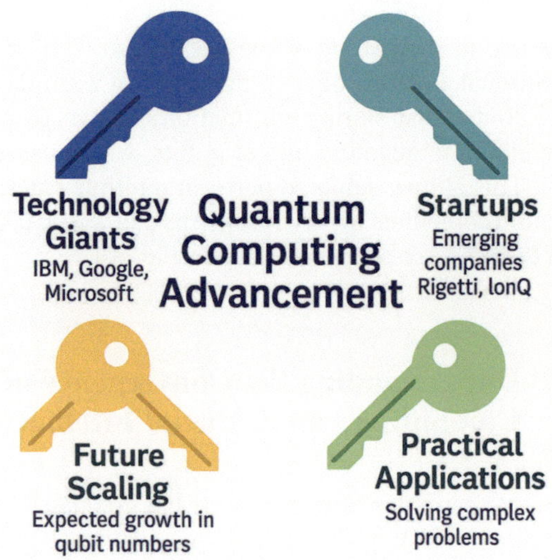

Fig. 8.2 Development of quantum computing

enable computers to learn from data, perceive their environment, make predictions, and act toward specific goals. AI is powering everything from virtual assistants like Siri and Alexa to Recommendation systems on Netflix and Amazon to self-driving cars by Tesla and Waymo [5].

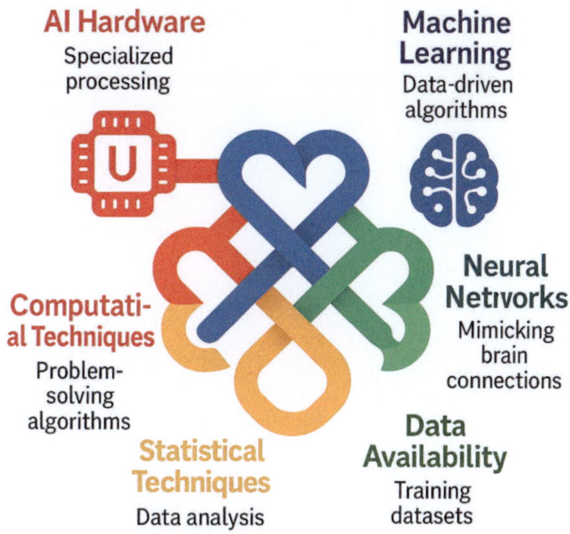

Fig. 8.3 Components of AI advancements

The meteoric rise of AI has been fueled by the availability of vast troves of data, advances in statistical and computational techniques, and the massively parallel processing capabilities of powerful specialized AI hardware like GPUs and TPUs. While highly potent, AI systems today are still narrow or "weak" AI, focused on performing specific tasks using predefined algorithms rather than achieving accurate human-level intelligence and versatility [6] as represented in Fig. 8.3.

Quantum AI refers to the intersection of quantum computing and AI. This merger promises to unleash a new paradigm of computing that combines the exponential scale and multidimensional processing capacities of quantum systems with the self-learning intelligence of AI [7]. Quantum machine learning techniques are being developed to run AI algorithms on quantum devices. The strange properties of qubits could allow AI systems to perceive, learn, decide, and optimize in radically new ways.

On a fundamental level, quantum AI can overcome limitations ingrained in classical AI systems. Traditional AI relies on crunching through data using classical hardware like CPUs and GPUs. Classical processing hardware's linear, binary nature restricts how AI models perceive data relationships and limits their performance [8]. By exploiting the ability of quantum systems to process information in multiple states and directions simultaneously, quantum AI could shatter these constraints and create exponentially more powerful AI as depicted in Fig. 8.4.

Some promising applications that could emerge from this synergy of quantum computing and AI [9] include as shown in Fig. 8.5.

- *Quantum Neural Networks*—Existing neural network architectures could be ported to quantum platforms, enabling dramatically faster training and more predictive capabilities. Quantum neural networks could provide exponential increases in speed and scalability.

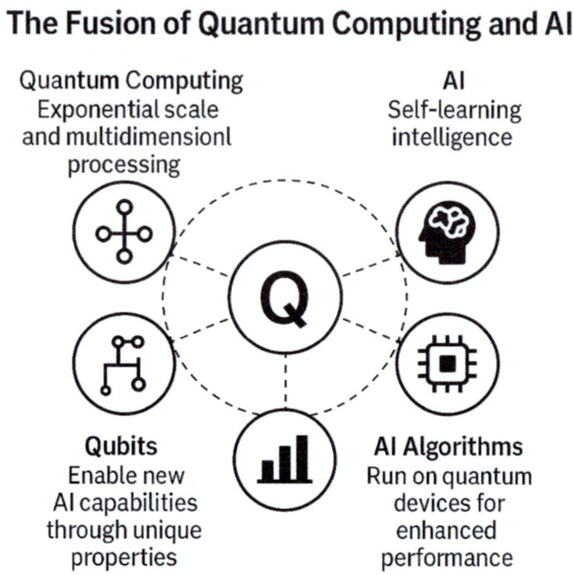

Fig. 8.4 Fusion of quantum computing and artificial intelligence

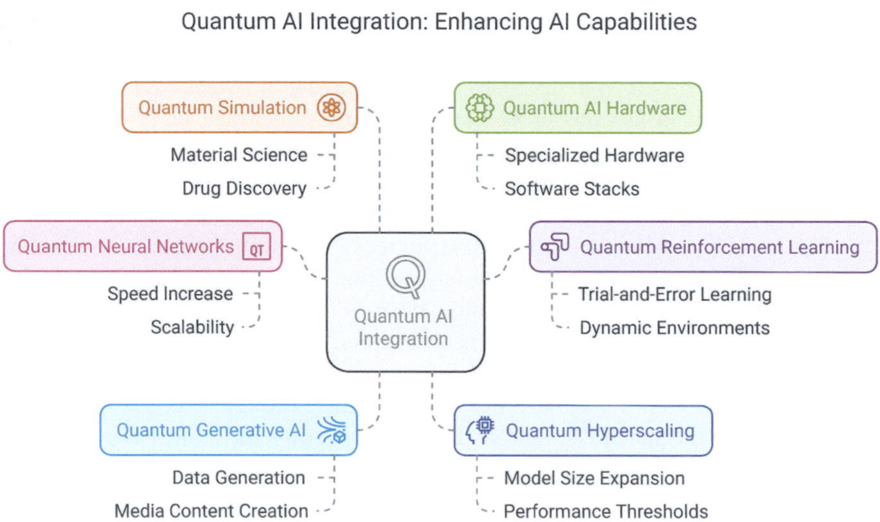

Fig. 8.5 Quantum AI integration

- *Quantum Reinforcement Learning*—Reinforcement learning agents could utilize quantum processors to radically accelerate learning processes based on trial-and-error interactions with dynamic environments.
- *Quantum Generative AI*—Combining quantum circuits with generative adversarial networks and other generative models could create AI systems that

more efficiently generate synthetic data, media content, molecular structures, pharmaceutical drugs, and more.

- *Quantum Hyperscaling*—Processing tensor networks on quantum systems may allow the hyperscaling of AI models beyond current size and performance thresholds. This could enable more powerful CNNs, Transformers, and other leading AI model architectures.
- *Quantum Simulation*—Quantum AI systems could perform high-fidelity simulations of molecular, atomic, and subatomic phenomena, leading to breakthroughs in material science, drug discovery, industrial chemistry, and other fields.
- *Quantum AI Hardware*—Specialized quantum hardware and software stacks optimized to run AI algorithms and models at scale.

The business and economic potential of this quantum AI convergence is astounding. IDC predicts that by 2028, 25% of all new enterprise IT solutions will run on quantum algorithms, accounting for $2.2 trillion in global GDP impact. In the next decade, quantum AI could become the most potent technology asset that enterprises can leverage to transform everything from financial risk modeling to supply chain optimization. Management professionals who understand and embrace this emerging paradigm will gain a competitive advantage.

8.2 Relevance of Quantum AI for Modern Management

Quantum AI has the potential to transition management science into a new era. Managers could utilize quantum AI capabilities to completely transform how they model uncertainty, identify insights, formulate strategies, and optimize complex systems [10] represents in Fig. 8.6.

Some of the critical areas where quantum AI can impact management practices include:

1. ***Risk Management***—Quantum AI could enable more accurate risk profiling and uncertainty modeling for everything from financial portfolios to project management and strategic planning. By processing probabilistic scenarios in multidimensional Hilbert spaces, quantum risk models may find correlations that classical models cannot [11].
2. ***Logistics and Supply Chain Optimization***—Quantum optimization algorithms can enable breakthroughs in routing delivery fleets, inventory management, and dynamic pricing. Aerospace companies like Airbus have already applied quantum computing to optimize manufacturing and maintenance processes [12].
3. ***Predictive Analytics***—Quantum neural networks and quantum-assisted machine learning models may uncover novel signals and insights for predictive analytics across diverse data sets, such as financial, meteorological, manufacturing, IoT, and e-commerce data [13].
4. ***Cybersecurity***—Quantum AI has promising applications in cybersecurity, cryptography, and fraud detection. Quantum software can analyze larger volumes of

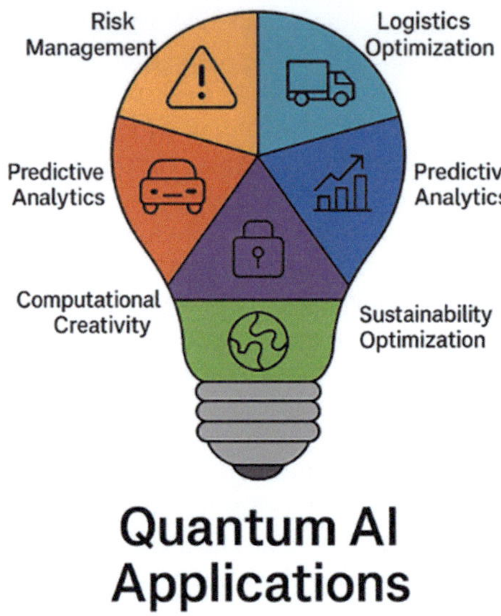

Fig. 8.6 Quantum artificial intelligence applications

network traffic data to identify threats. Post-quantum encryption schemes will also be essential to protect systems from quantum cyberattacks [14].

5. ***Autonomous Systems***—Quantum processors may allow autonomous vehicles, robots, and drones to make navigation decisions faster and optimize trajectories by simultaneously processing spatial, visual, and sensor data. Volkswagen, Daimler, and other automobile giants are already exploring quantum computing R&D initiatives [15].
6. ***Computational Creativity***—Quantum generative models like quantum GANs could enhance computer-aided creative tasks like generating new products, services, and business concepts and strategies. Quantum AI systems may also create original art, music, architecture, and design concepts [16].
7. ***Sustainability and ESG Optimization***—Quantum optimization and simulation can enable more rapid advances in sustainability across operations, manufacturing, energy systems, transportation networks, and value chains. Quantum analytics may derive hidden ESG (environmental, social, and governance) insights from sustainability data [17].
8. ***Quantum of Business***—Combining quantum behavioral science with quantum AI could lead to breakthroughs in behavioral economics, finance, organizational psychology, and other social science domains, unlocking new insights into human decision-making for marketing, policymaking, and management [18].

Managers leveraging quantum AI should identify high-value business cases where quantum capabilities can drive competitive differentiation. This involves assessing where classical analytics and computing techniques are hitting limitations today and recognizing how quantum AI's exponential scale, multidimensional data processing and predictive capabilities can shatter these constraints. Management should also invest in quantum literacy and skills development across their organizations to fully capitalize on quantum AI's potential as it matures over the next decade.

Early proof-of-concept projects, research partnerships with quantum computing companies, building in-house skills, and initiating C-suite discussions on long-term quantum strategy will be prudent first steps for management teams looking to navigate the quantum AI revolution. Change management will also be critical, as quantum AI may significantly disrupt existing business models, processes, and roles. Overall, forward-thinking managers who take steps today to embrace quantum AI will be firmly positioned to amplify their competitive advantage and shape the quantum future of their industries.

References

1. McKinsey & Company. *Quantum computing comes of age.* Accessed 12 Feb 2023. https://www.mckinsey.com/business-functions/mckinsey-digital/our-insights/quantum-computing-comes-of-age
2. Deloitte Insights. *Quantum computing powers up artificial intelligence.* Accessed 12 Feb 2023. https://www2.deloitte.com/us/en/insights/industry/technology/technology-media-and-telecom-predictions/2022/quantum-computing-powers-up-artificial-intelligence.html
3. IBM. *Quantum for business leaders.* Accessed 12 Feb 2023. https://www.ibm.com/quantum-computing/learn/what-is-quantum-computing/
4. Rigetti. *What is quantum AI*? Accessed 12 Feb 2023.
5. Bhoumik, D., Sur-Kolay, S., Kumar K. J. L., & Iyengar, S. S. (2023). *Synergy of machine learning with quantum computing and communication.* arXiv.
6. Melnikov, A., Kordzanganeh, M., Alodjants, A., & Lee, R.-K. (2023). *Quantum machine learning: From physics to software engineering.* arXiv.
7. Krenn, M., Landgraf, J., Foesel, T., & Marquardt, F. (2022). *Artificial intelligence and machine learning for quantum technologies.* arXiv.
8. Garg, S., & Ramakrishnan, G. (2020). *Advances in quantum deep learning: An overview.* arXiv.
9. Turner, P., & Wittek, P. (2018). Machine learning & artificial intelligence in the quantum domain: A review of recent progress. *Reports on Progress in Physics, 81*(7).
10. Schuld, M., Sinayskiy, I., & Petruccione, F. (2014). *Supervised learning with quantum computers.* Springer.
11. Biamonte, J., Wittek, P., Pancotti, N., Rebentrost, P., & Wiebe, N. (2017). Quantum machine learning. *Nature.*
12. Pasqal. (2024–2025). *Quantum AI explained: New guide for leaders ready to innovate.*
13. Moody's. (2024–2025). *The next frontier after AI agents: Quantum artificial intelligence.* Moody's Insights.
14. AWS. (2025). *What is quantum AI?* AWS Educational Resource.
15. GDIT. (2023). *The next AI wave: Quantum AI.* GDIT Perspectives.
16. Optionstrading.org. (2024–2025). *Quantum AI trading: Impact on options.*
17. Tristar Software. (2024). *How is quantum AI transforming investment strategies.*
18. Forbes. (2024). *Quantum AI: What good is it?* Forbes Business Council.

Chapter 9
Quantum Machine Learning: A Primer

9.1 Quantum Versus Classical Machine Learning

Classical machine learning has become ubiquitous for practical pattern recognition and predictive analytics tasks, powering applications from image classification to stock market forecasting. Machine learning models like neural networks can make predictions and decisions without relying on predefined rigid program logic by learning from sample data [1]. However, training these models entails intense computational requirements as shown in Fig. 9.1.

For example, deep neural networks can take weeks to train on powerful GPU clusters using optimization techniques like gradient descent. The computations involved do not require deep reasoning but rather repeated matrix manipulation steps like multiplication and inversion [2]. Quantum computing promises a route to accelerate these brute-force yet essential computations underpinning machine learning as represented in Fig. 9.2.

Quantum machine learning explores how quantum mechanical phenomena can speed up learning tasks to enable otherwise intractable model training. It aims to benefit from quantum parallelism, which allows quantum systems to simultaneously evaluate functions across many inputs. This parallel processing capacity can provide polynomial or exponential speedups over classical methods [3].

While still nascent, quantum machine learning algorithms are rapidly being developed. Leading application areas include pattern recognition, optimization, sampling, quantum neural networks, parameter learning, quantum-enhanced dimensionality reduction, and quantum reinforcement learning [4]. Proof-of-concept demonstrations have shown order-of-magnitude runtime improvements on benchmark tests using quantum devices and simulators as represented in Fig. 9.3.

However, translating these promising theoretical speedups into practical business advantages faces formidable technical obstacles. Noise and decoherence effects plague existing quantum processors, restricting them to small-scale computations. New algorithmic and software advances are also required to map machine learning

M. Singh and B. S. Rawal, *Quantum Minds: Exploring Quantum Computing and AI Synergy*, https://doi.org/10.1007/978-981-95-2498-3_9

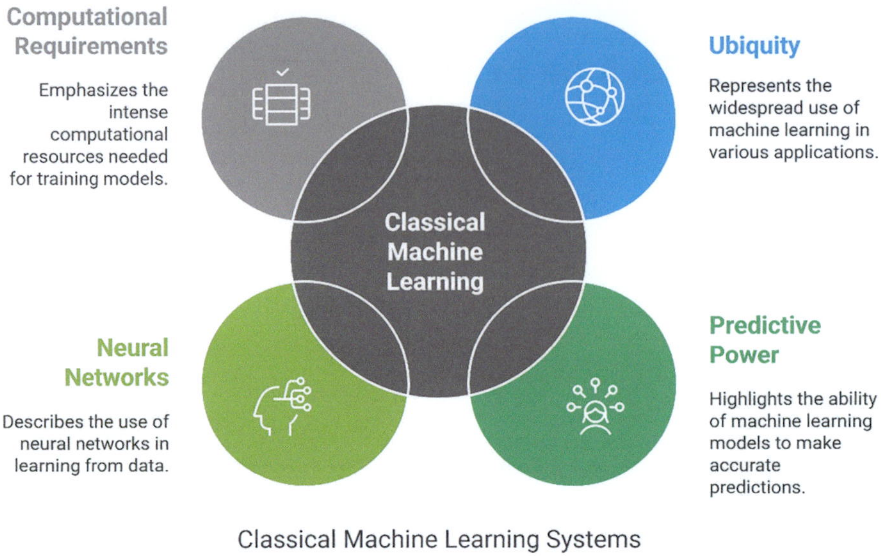

Fig. 9.1 Classical machine learning system

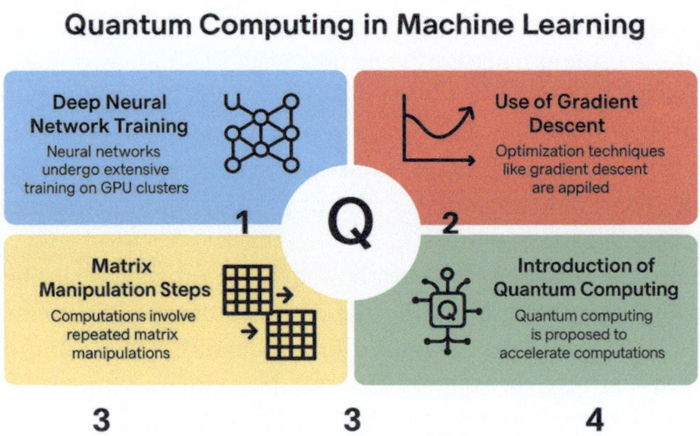

Fig. 9.2 Quantum computing in machine learning system

programs onto quantum hardware optimally. Nonetheless, quantum machine learning provides a vital bridge between quantum computing and artificial intelligence and will enable AI systems to tap into the power of quantum information processing [5].

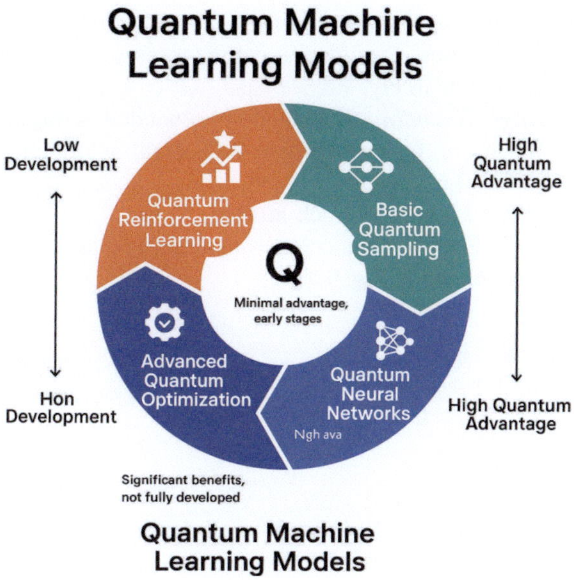

Fig. 9.3 Quantum machine learning models

9.2 Quantum Data Encoding and Representation

Classical machine learning represents data using arrays of binary numbers. Quantum ML uses qubits, which can explore correlations across data by encoding information in superposition states as shown in Fig. 9.4.

Key representations include:

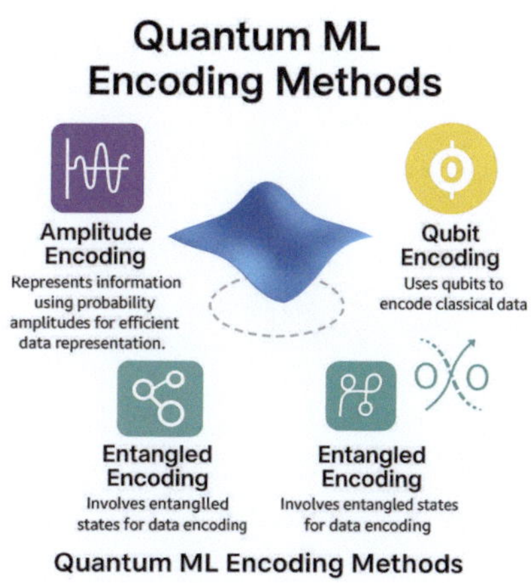

Fig. 9.4 Quantum machine learning encoding methods

- ***Amplitude Encoding***: Information is encoded in probability amplitudes of qubit basis states. Provides a natural representation for probabilities/intensities as represents in Fig. 9.5.
- ***Angle Encoding***: Information is encoded in relative phases between basis states. It is useful for representing values with periodicity, like angles or time as shown in Fig. 9.6.
- ***Qubit Encoding***: Each attribute value is represented by a corresponding qubit state. For example, color = red could be $|0\rangle$, color = green could be $|1\rangle$, etc. as represented in Fig. 9.7.

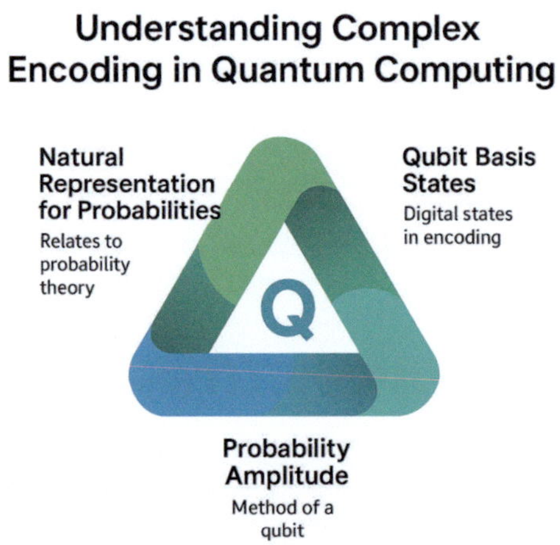

Fig. 9.5 Amplitude encoding in quantum computing

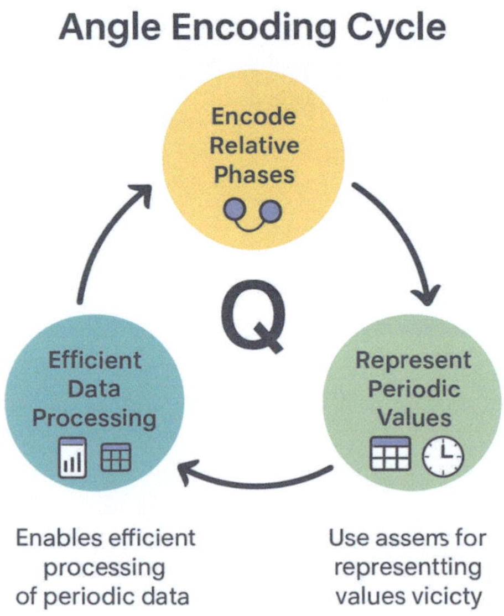

Fig. 9.6 Amplitude encoding cycle

Fig. 9.7 Qubit encoding cycle

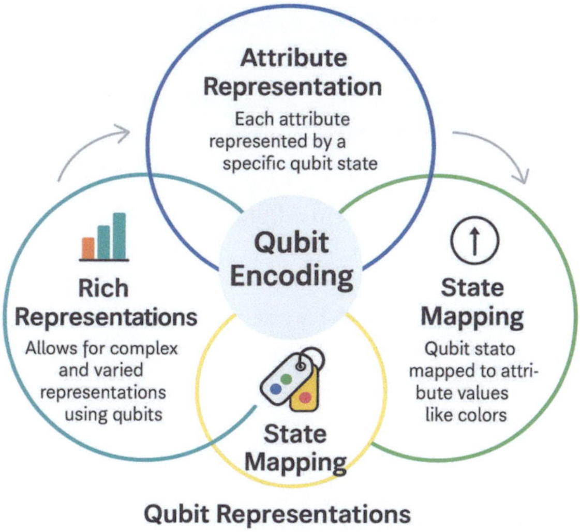

- ***Entangled Encoding***: Information can be encoded in entangled states across qubits, enabling quantum-enhanced dimensionality reduction strategies.

These quantum state representation schemes equip quantum ML models to uncover insights that are difficult for classical ML with simple binary data encoding. However, encoding real-world datasets into quantum states remains challenging [6]. Hybrid strategies are emerging, where classical preprocessing converts datasets into formats amenable to quantum processing as represented in Fig. 9.8.

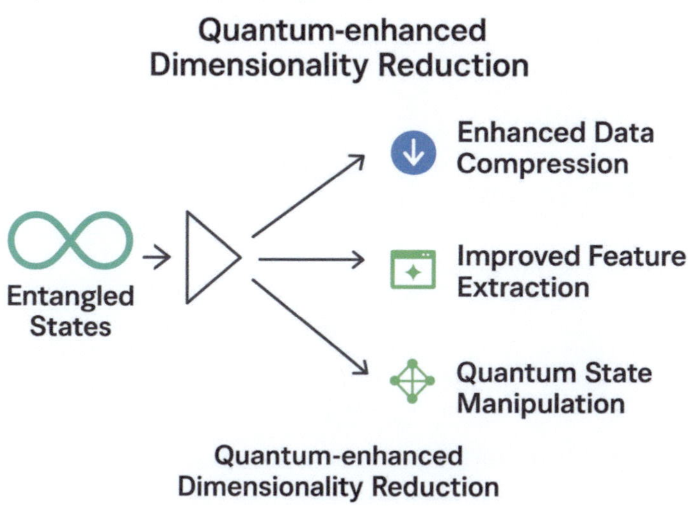

Fig. 9.8 Entangled encoding cycle

9.3 Quantum Supervised Learning Algorithms

Supervised learning uses labeled datasets to train predictive models like classifiers or regressors as represents in Fig. 9.9.

Here are some promising quantum ML algorithms for supervised learning tasks:

Quantum Classification Algorithms

See Fig. 9.10.

- *Quantum Neural Networks*: Quantum computational units replace classical neurons. Can enable exponential speedups in training via parallel weight updates as represented in Fig. 9.11.
- *Quantum Decision Trees*: Traverse decision trees superposition across multiple paths to accelerate identifying optimal splits as represented in Fig. 9.12.

Quantum Regression Algorithms

See Fig. 9.13.

- *Quantum Support Vector Machine*: This machine encodes training data in quantum states and leverages quantum matrix inversion and phase estimation to rapidly find the optimal separatrix between classes as represented in Fig. 9.14.
- *Quantum linear regression (QLR) Encodes training data in quantum states and* leverages quantum matrix inversion and eigenvalue estimation to rapidly fit predictive models as shown in Fig. 9.15.

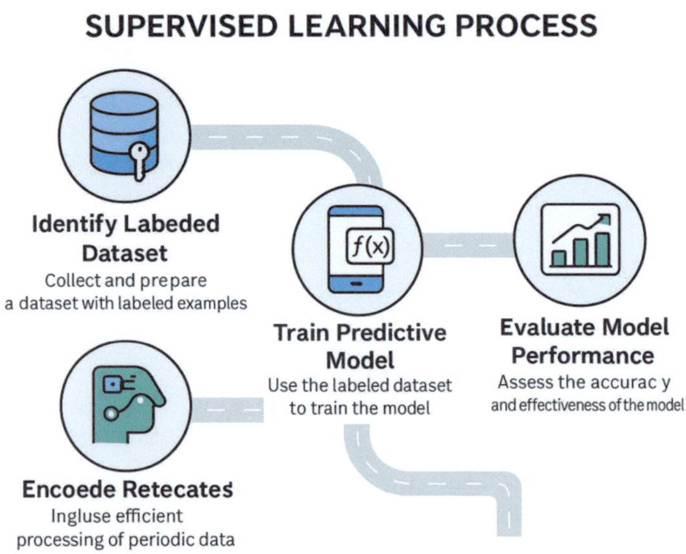

Fig. 9.9 Quantum supervised learning algorithms

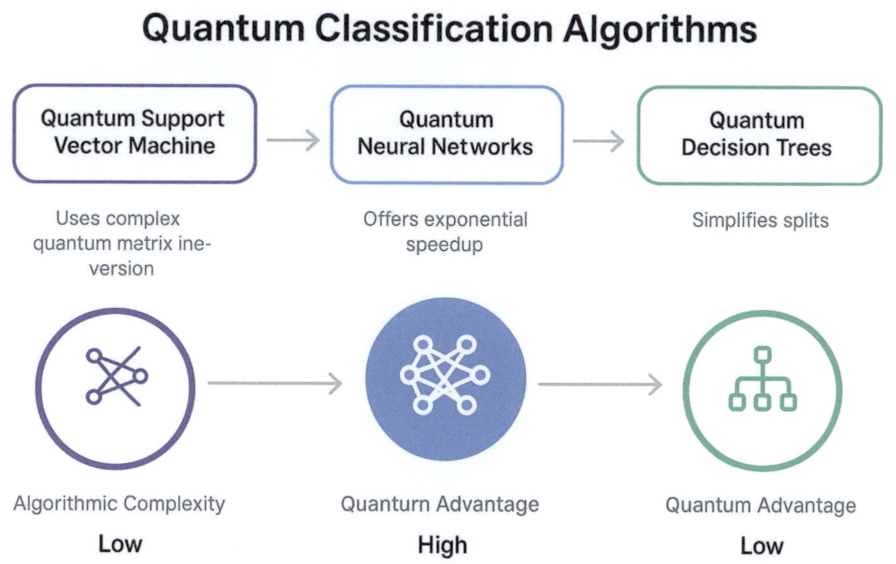

Fig. 9.10 Quantum classification models

Fig. 9.11 Quantum neural networks

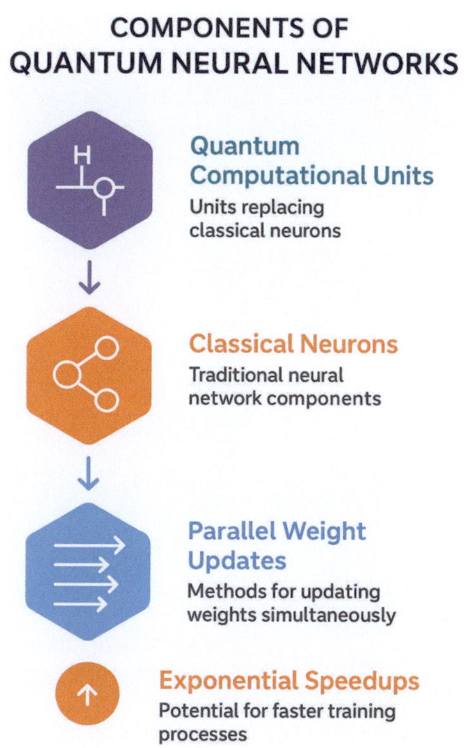

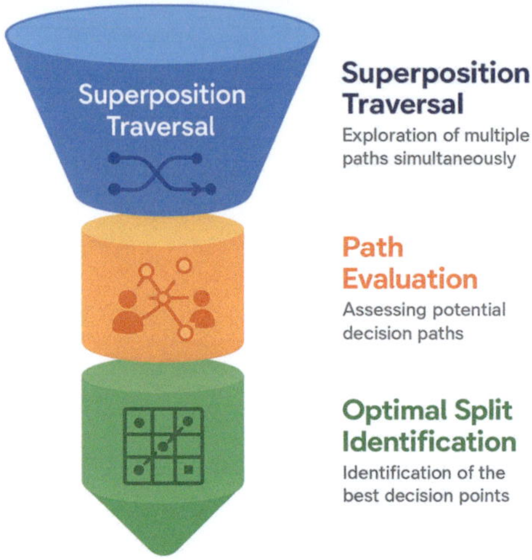

Fig. 9.12 Quantum decision trees

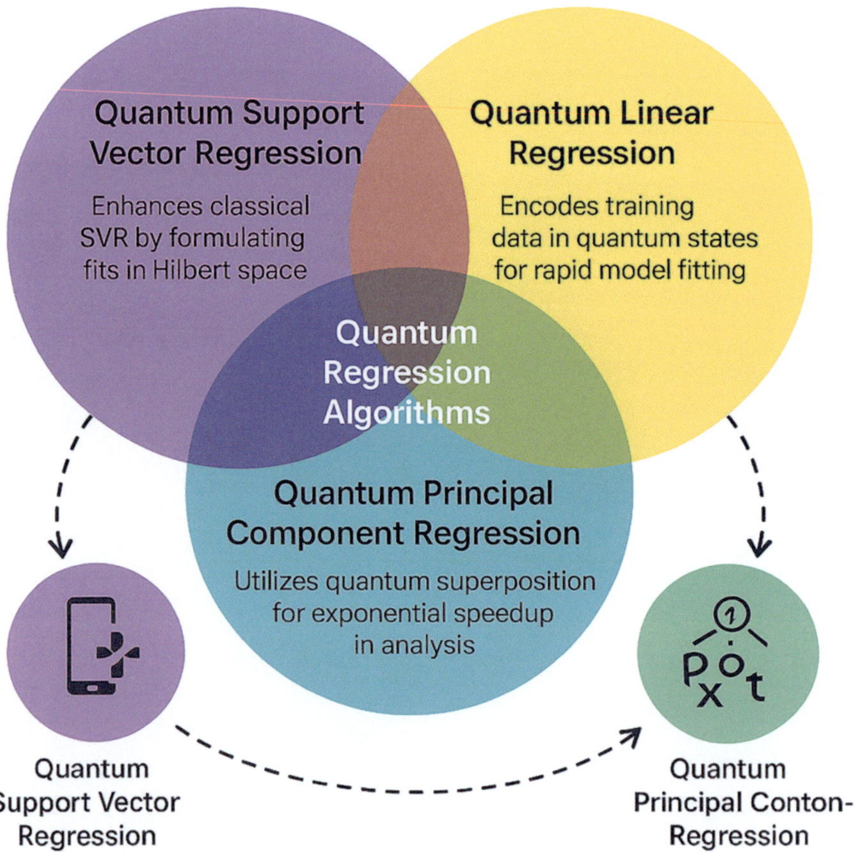

Fig. 9.13 Quantum regression algorithms

Fig. 9.14 Quantum support vector machine learning algorithm

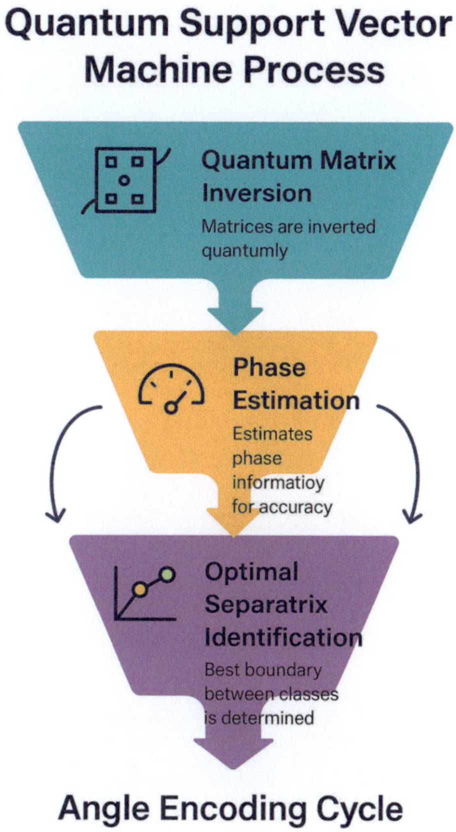

- *Quantum Principal Component Regression*: Performs principal component analysis thoroughly in quantum superposition to enable exponential speedup as depicted in Fig. 9.16.
- *Quantum Support Vector Regression*: Quantum enhancement of classical SVR method by formulating optimal fit in Hilbert space as represented in Fig. 9.17.

Quantum Unsupervised Learning Algorithms

Unsupervised learning analyzes unlabeled datasets to uncover intrinsic patterns, clusters, and associations as shown in Fig. 9.18.

Quantum-enhanced techniques include:

- *Quantum Clustering*: Rapidly identifies clusters in data by superposition across trial cluster configurations and parallel evaluation of objective functions like inertia or silhouette coefficients as Fig. 9.19.

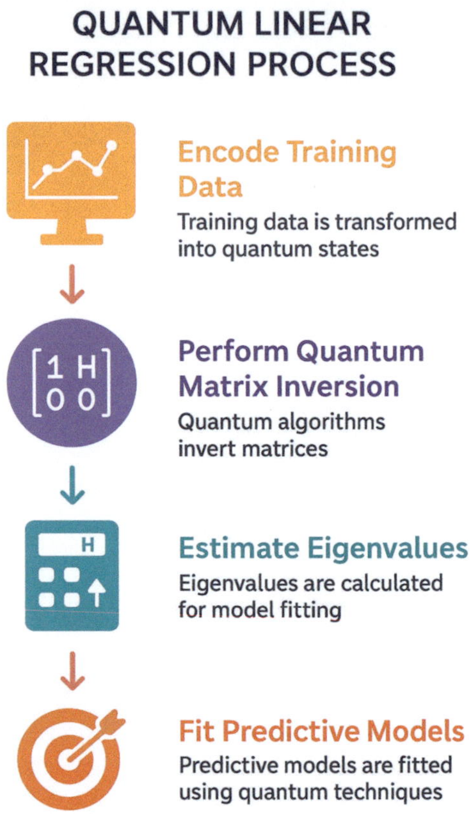

Fig. 9.15 Quantum linear regression

- *Quantum Dimensionality Reduction*: Projects data into lower dimensional spaces for analysis using quantum embeddings derived via quantum matrix manipulation as represented in Fig. 9.20.
- *Quantum Anomaly Detection*: Identifies anomalies and outliers by simultaneously comparing the distance between quantum encoded data points as shown in Fig. 9.21.
- *Quantum Generative Adversarial Networks*: Enables more stable training of GANs by encoding latent variables and data points in quantum states as represented in Fig. 9.22.

Quantum Reinforcement Learning

Reinforcement Learning (RL) develops autonomous agents that optimize behavior by repeated trial-and-error interactions with dynamic environments as depicted in Fig. 9.23.

Quantum techniques can accelerate RL by faster training of agents.

Fig. 9.16 Quantum principal component regression

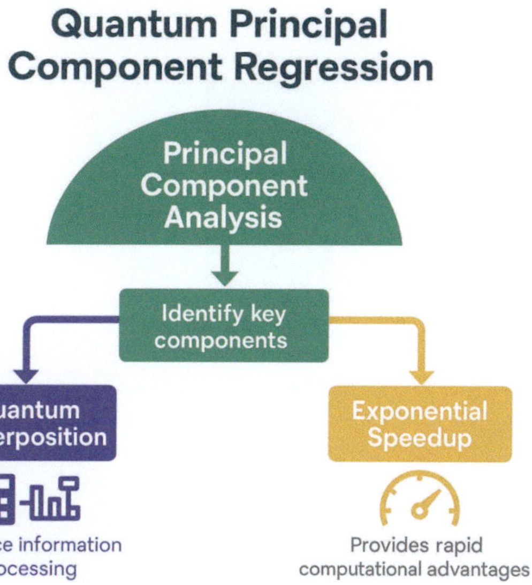

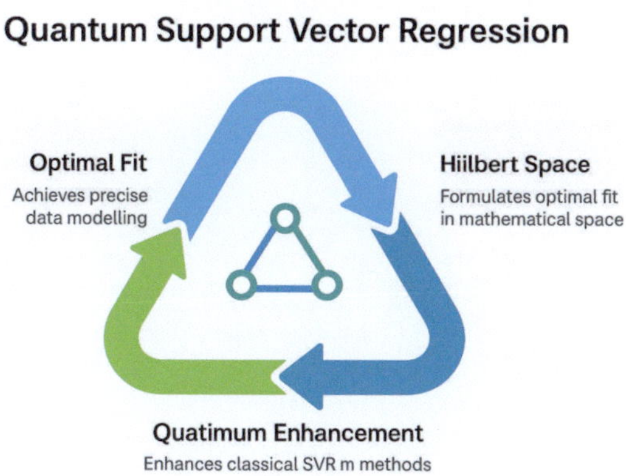

Fig. 9.17 Quantum support vector regression

- *Quantum Q-Learning*: Updates Q-values across multiple states and actions simultaneously by amplitude amplification, enabling exponential speedup in training agents as represented in Fig. 9.24.
- *Quantum Policy Gradients*: Leverages quantum gradient computation across multiple trajectories to speed up optimizing policies as shown in Fig. 9.25.
- *Quantum Value Iteration*: Uses Grover's algorithm to quadratically speed up dynamic programming in Markov decision processes as represented in Fig. 9.26.
- *Quantum Monte Carlo*: Generates probabilistic returns samples by quantum random walks to speed up policy evaluation as shown in Fig. 9.27.

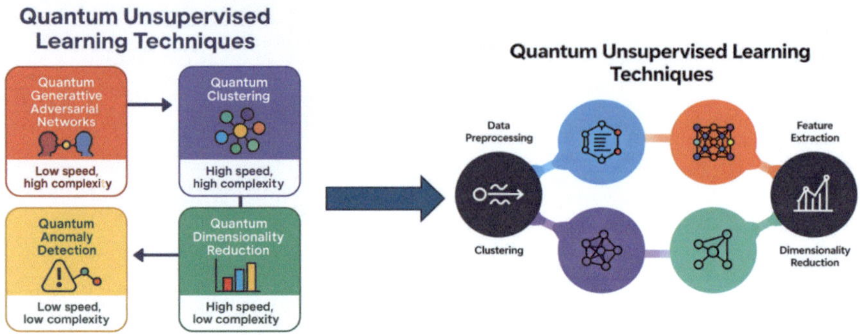

Fig. 9.18 Quantum unsupervised learning models

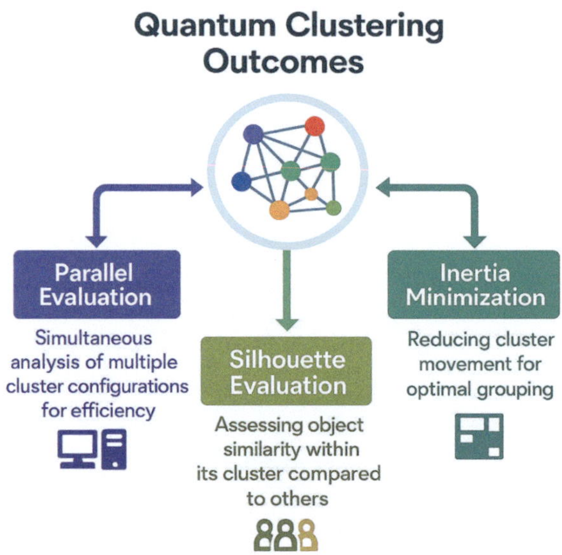

Fig. 9.19 Quantum clustering outcomes

Summary

Quantum machine learning opens exciting possibilities at the intersection of artificial intelligence and quantum information processing. Early theoretical progress shows promising runtime advantages over classical methods for supervised, unsupervised, and reinforcement learning tasks. However, multiple technical challenges remain in translating these speedups into practical business advantages. Building quantum ML workflows that integrate seamlessly with existing classical ML pipelines will be essential for adoption. Quantum machine learning provides a glimpse into the shift from artificial intelligence to quantum intelligence.

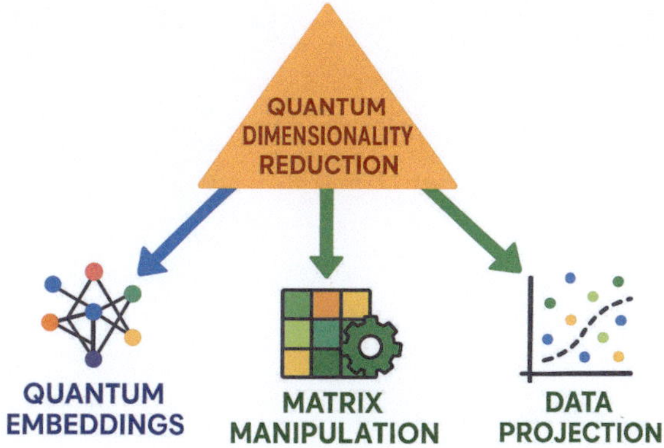

Fig. 9.20 Quantum dimensionality reduction

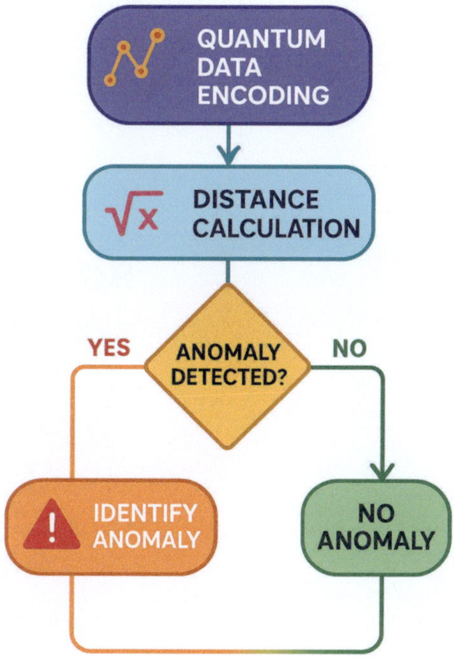

Fig. 9.21 Quantum anomaly detection

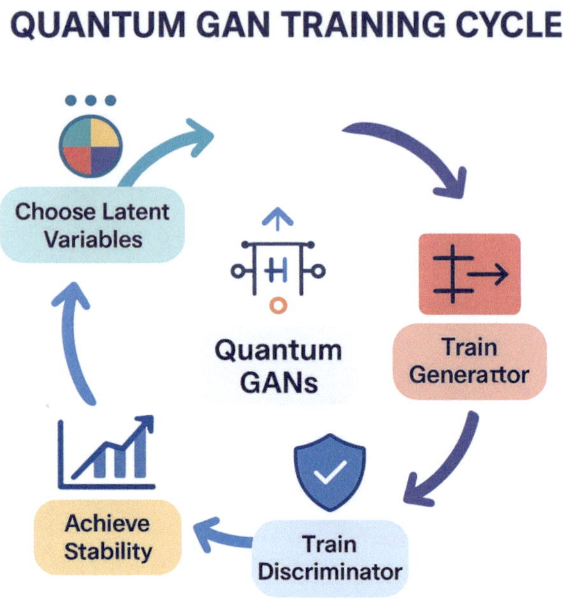

Fig. 9.22 Quantum generative adversarial networks

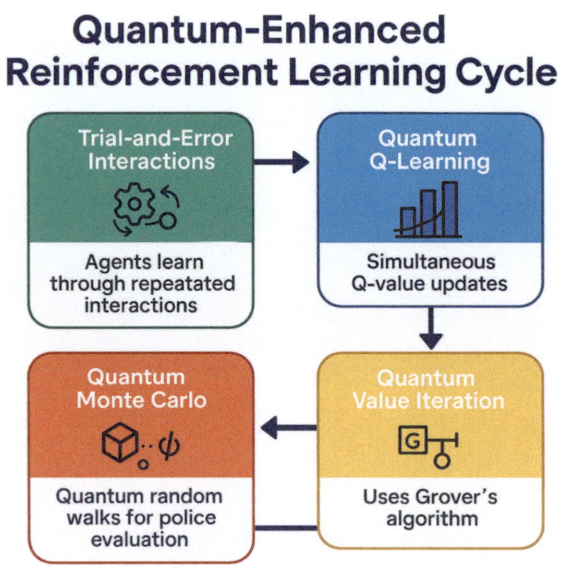

Fig. 9.23 Quantum reinforcement learning

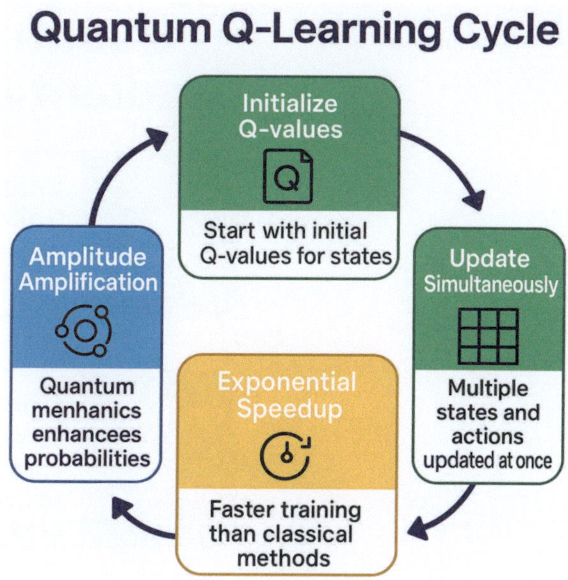

Fig. 9.24 Quantum Q-Learning cycle

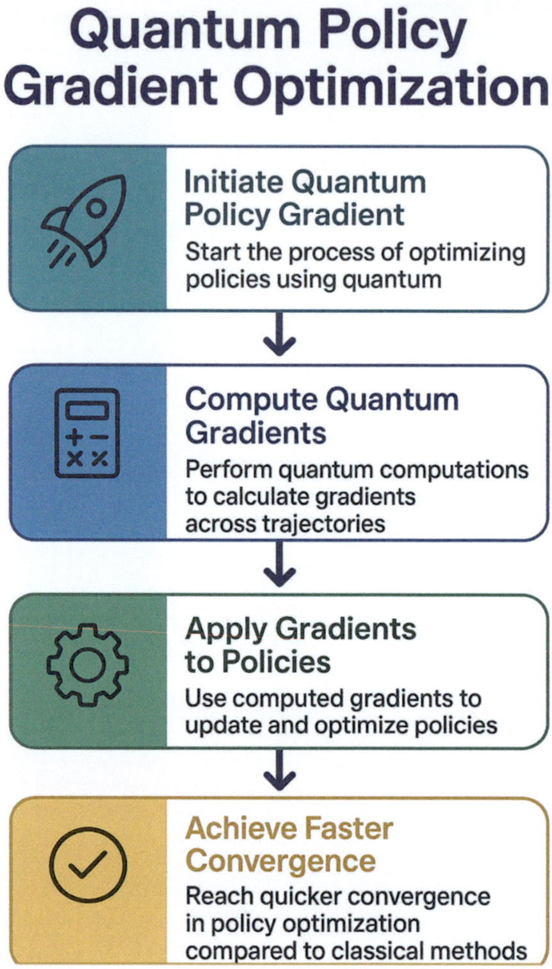

Fig. 9.25 Quantum policy gradient optimization

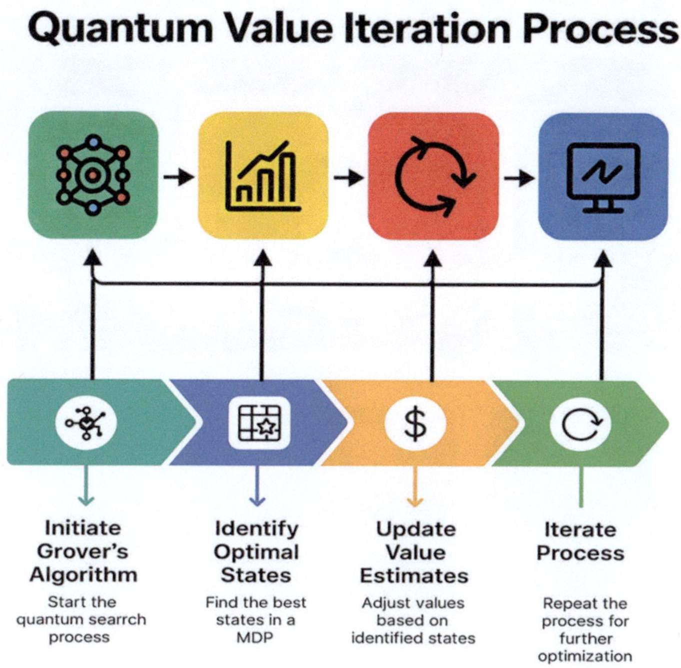

Fig. 9.26 Quantum value iteration

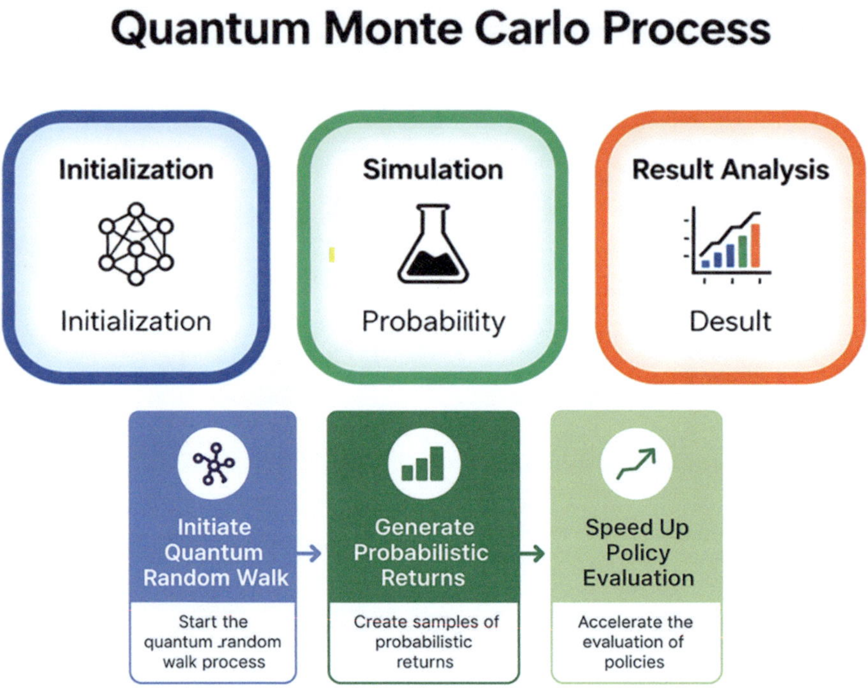

Fig. 9.27 Quantum Monte Carlo

References

1. Wang, Y., & Liu, J. (2024). *A comprehensive review of quantum machine learning: From NISQ to fault tolerance.* arXiv.
2. Mishra, N., Kapil, M., Rakesh, H., Anand, A., et al. (2021). *Quantum machine learning: A review and current status.* arXiv.
3. Benedetti, M., Lloyd, E., Sack, S., & Fiorentini, M. (2019). *Parameterized quantum circuits as machine learning models.* arXiv.
4. Dunjko, V., & Briegel, H. J. (2018). Machine learning & artificial intelligence in the quantum domain: A review of recent progress. *Reports on Progress in Physics, 81*(7), 074001.
5. Peddireddy, D., Bansal, V., Jacob, Z., & Aggarwal, V. (2022). Tensor ring parametrized variational quantum circuits for large-scale quantum machine learning. *NPJ Quantum Information.*
6. Shrivastava, A. & Sharma, V. (2024). Primer on data in quantum machine learning. In *Proceedings of ESWEEK 2024.*

Chapter 10
Quantum Enhancements for AI Models

The previous section provided a broad overview of quantum machine learning techniques. This section dives deeper into quantum enhancements for specific AI models like neural networks, transformers, and reinforcement learning agents. Early research shows that integrating quantum processing into classical AI algorithms can speed training, increase model capacity, and improve predictive accuracy. However, practical implementation remains challenging [1]. This section equips readers with conceptual foundations on quantum augmentation approaches for leading AI models today and highlights opportunities and limitations as represented in Fig. 10.1.

10.1 Quantum Data Processing and Dimensionality Reduction

Real-world datasets for training AI models typically consist of large numbers of interconnected attributes across millions of samples. Processing this high-dimensional data often requires simplification before further analysis. Quantum computing provides faster dimensionality reduction techniques by exploiting quantum superposition to analyze multidimensional data representations efficiently [3] as represents in Fig. 10.2.

Quantum Principal Component Analysis (QPCA) can efficiently derive important patterns in large datasets. PCA rotates correlated attributes into principal component axes using eigenvalue decomposition of the covariance matrix. QPCA performs this eigenanalysis thoroughly in quantum superposition using the QLA technique, enabling exponential speedups [3]. This allows faster dimensionality reduction as a preprocessing step as shown in Fig. 10.3.

M. Singh and B. S. Rawal, *Quantum Minds: Exploring Quantum Computing and AI Synergy*, https://doi.org/10.1007/978-981-95-2498-3_10

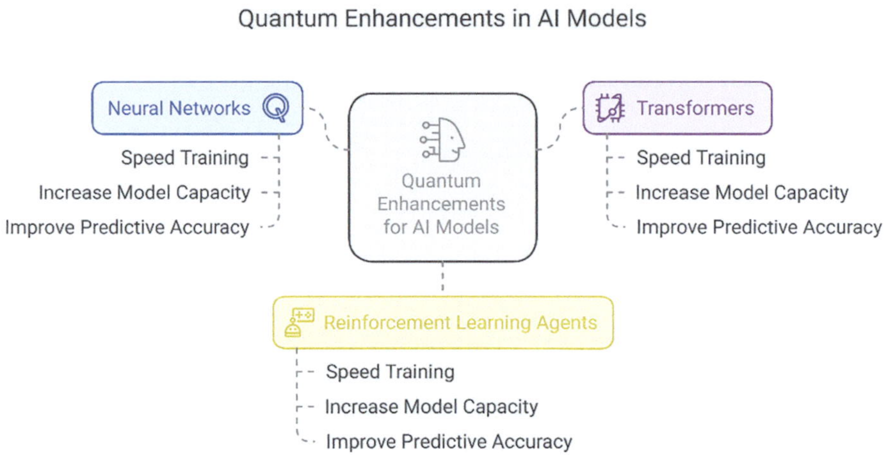

Fig. 10.1 Quantum enhancements in AI models

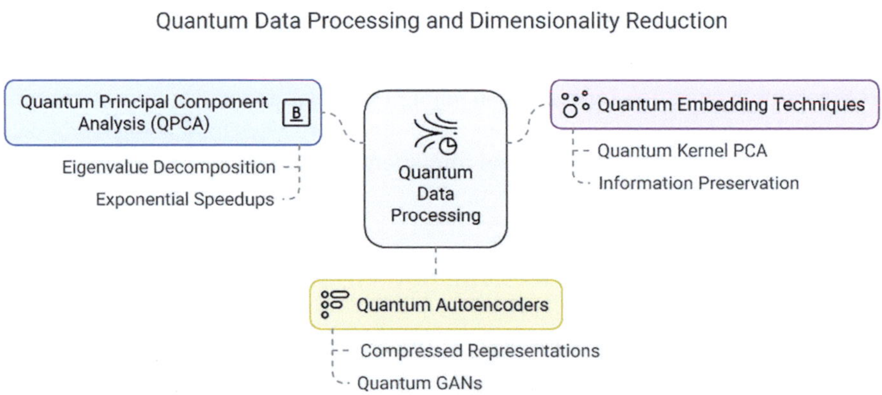

Fig. 10.2 Quantum data processing and dimensionality reduction

Quantum Embedding Techniques like quantum kernel PCA can project high-dimensional classical data into powerful quantum state encodings that preserve essential information [4]. Operations on these quantum embeddings can reveal insights that are challenging to uncover from classical embeddings as depicted in Fig. 10.4.

Quantum Autoencoders leverage quantum circuits to encode and decode data into compressed latent representations. Quantum generative models, like quantum GANs [5], also hold promise as density estimators for complex data distributions as shown in Fig. 10.5.

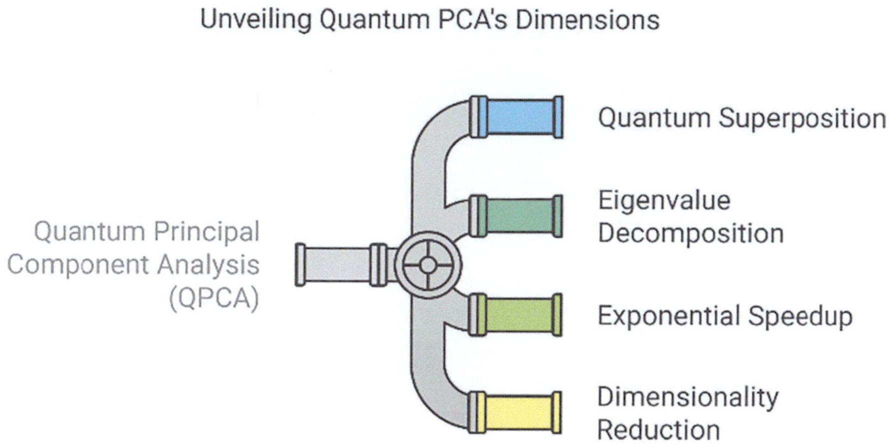

Fig. 10.3 Quantum principal component analysis (QPCA)

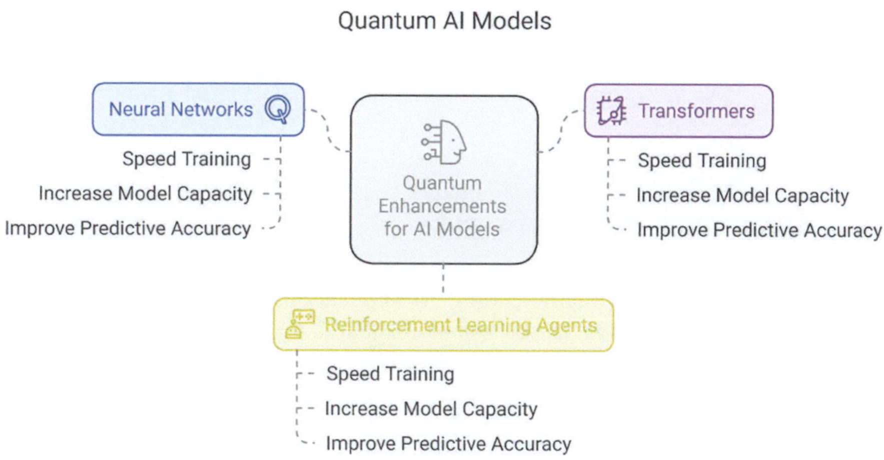

Fig. 10.4 Quantum AI models

Overall, quantum techniques allow AI models to tap into exponentially larger pools of information by projecting data into multidimensional Hilbert spaces. This enables deriving more powerful compressed representations than classical linear algebra approaches constrained by scalar matrix operations [7] as represents in Fig. 10.6.

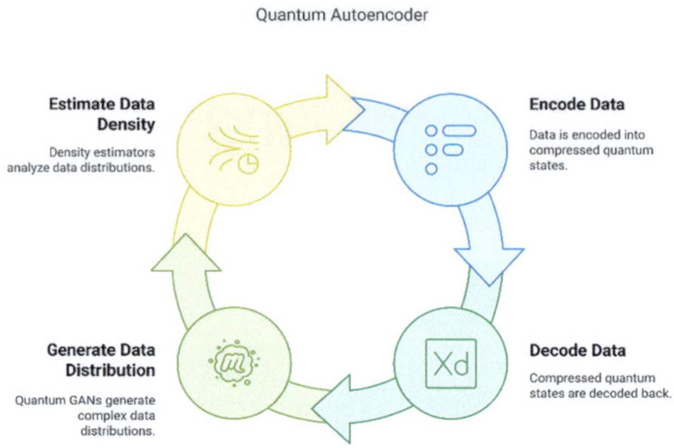

Fig. 10.5 Quantum autoencoders

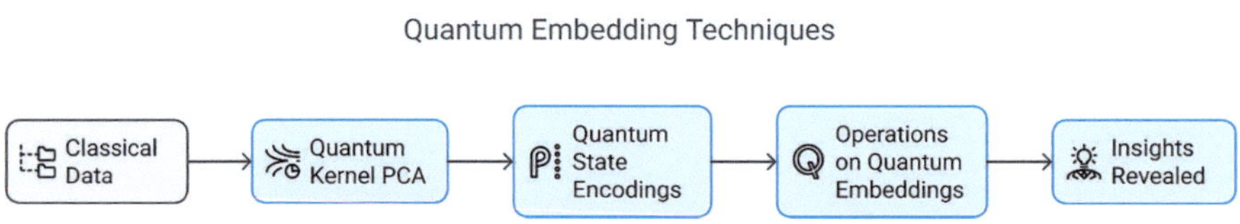

Fig. 10.6 Quantum embedding techniques

10.2 Quantum Neural Networks

Neural networks underpin many leading AI applications today, but their training remains computationally intensive. Quantum neural networks aim to accelerate training by replacing classical neurons with quantum computational units [8] (Fig. 10.7).

Different quantum neural network (QNN) architectures have been proposed:

- ***Quantum Perceptron's*** are basic neural units that leverage quantum gates for information aggregation and activation. Multi-layer networks of quantum perceptions can be constructed for representing highly complex functions as shown in Fig. 10.8.
- ***Quantum Tensor Neural Networks*** use quantum state embeddings for inputs rather than classical vectors. Logic operations are defined through qubit interactions [9]. Allows more robust learning for noisy data as depicted in Fig. 10.9.
- ***Quanvolutional Neural Network*** augment classical CNNs with Quan volution layers that act on quantum input states. This enables modeling quantum interactions and correlations in data (Fig. 10.10).

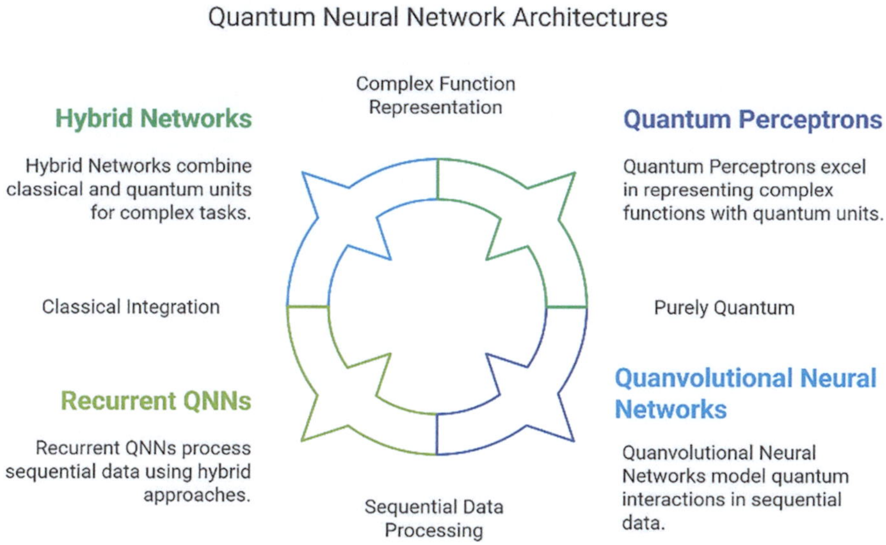

Fig. 10.7 Quantum neural network architectures

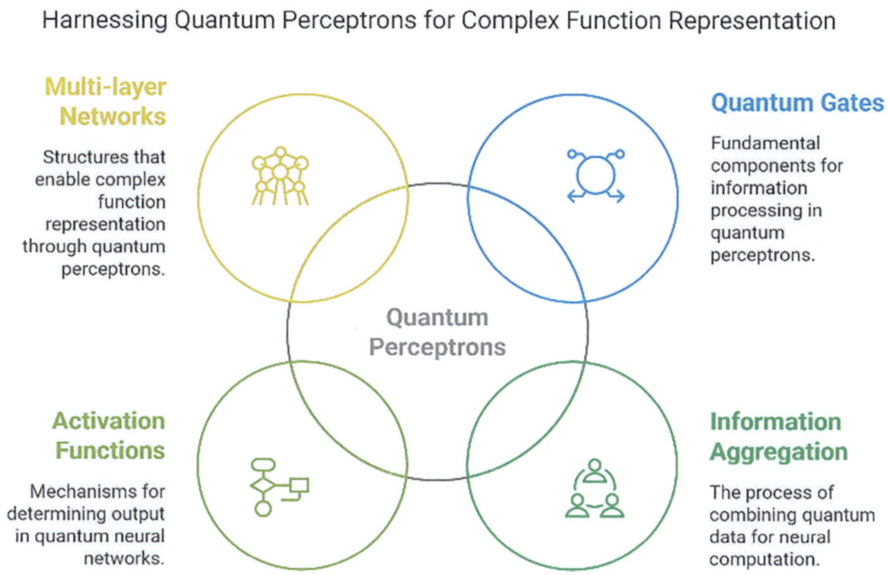

Fig. 10.8 Quantum perceptron's

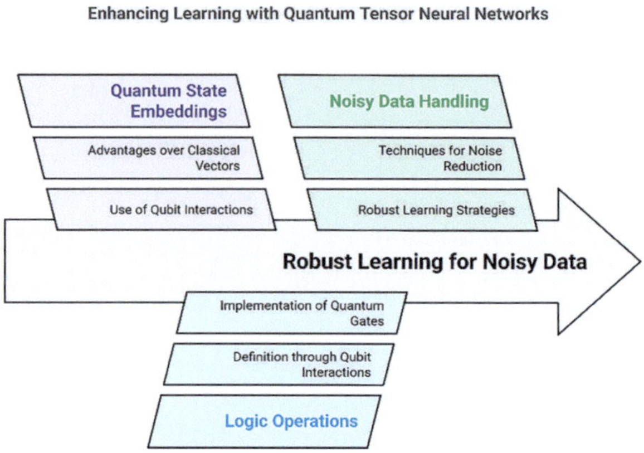

Fig. 10.9 Quantum tensor neural networks

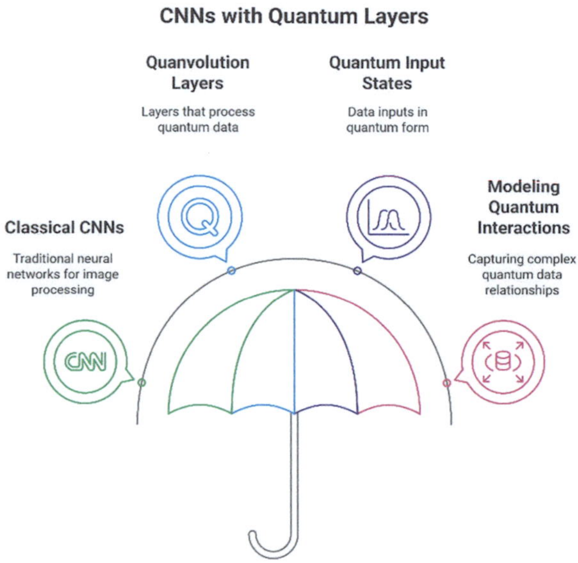

Fig. 10.10 Quanvolutional neural network

- ***Recurrent QNNs* utilize quantum RNN** architectures like Quantum LSTMs for processing sequential data like time series or natural language. Quantum recurrent models have longer memory retention (Fig. 10.11).
- ***Hybrid Networks*** leverage both classical and quantum units working in tandem to benefit from both processing paradigms as presented in Fig. 10.12.

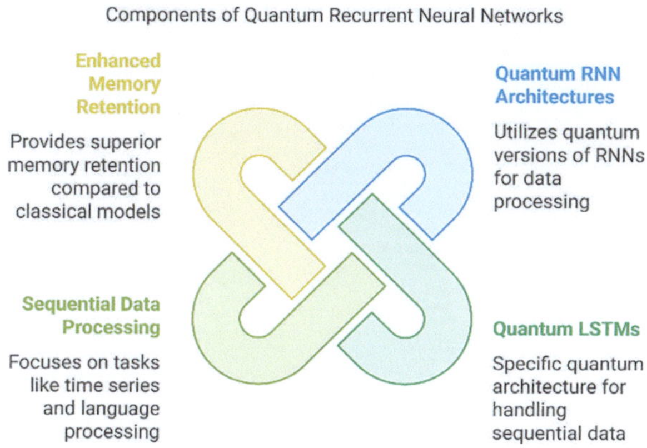

Fig. 10.11 Recurrent quanvolutional neural networks

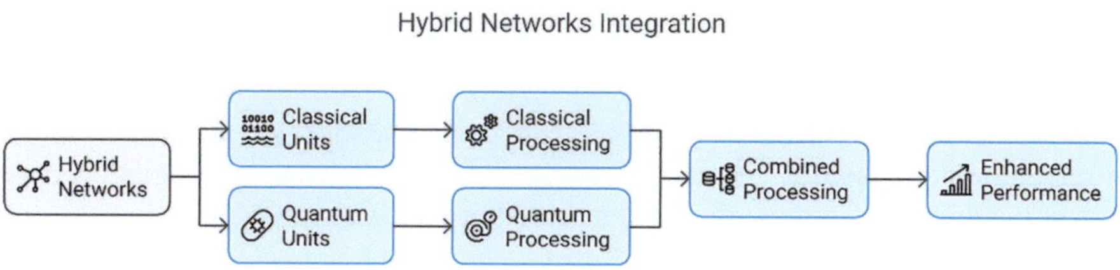

Fig. 10.12 Hybrid networks integration

The core training process of QNNs remains similar—adjusting synapse weights to minimize error. But quantum parallelism enables simultaneously evaluating error gradients across all weights and inputs in superposition. This allows orders of magnitude faster training compared to iterative classical backpropagation techniques [10].

However, realizing fully operational QNNs requires overcoming stability and scalability challenges with existing quantum hardware. Hybrid models combining classical and quantum units are more viable today [11]. Overall, QNNs promise to unlock unprecedented predictive power from exponentially larger datasets by exploiting quantum speedups.

10.3 Quantum Deep Learning

Deep learning using multi-layer neural networks underpins many cutting-edge AI solutions today, from computer vision to speech recognition. However, increasing model size and dataset complexity is straining even powerful classical HPC infrastructure as shown in Fig. 10.13.

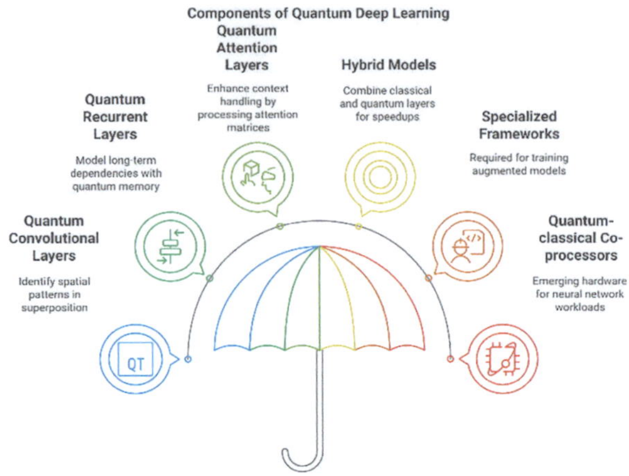

Fig. 10.13 Quantum deep learning

Quantum techniques offer a path to accelerate training and inference of large deep learning models.

Quantum Convolutional Layers in CNNs can identify spatial patterns across images in quantum superposition. Quanvolution operations on quantum image encoding extract features more efficiently as represented in Fig. 10.14.

Quantum Recurrent Layers in RNNs and LSTMs can model longer term temporal dependencies and sequences using quantum memory. This enhances processing natural language, time series, and genomics data as shown in Fig. 10.15.

Quantum Attention Layers enable smarter context handling for transformer networks by allowing quantum processing of the attention matrix. Reduces complexity for linking distant inputs as shown in Fig. 10.16.

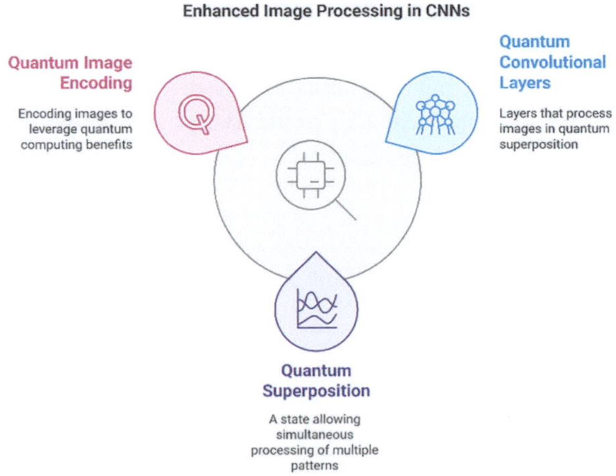

Fig. 10.14 Quantum convolutional layers

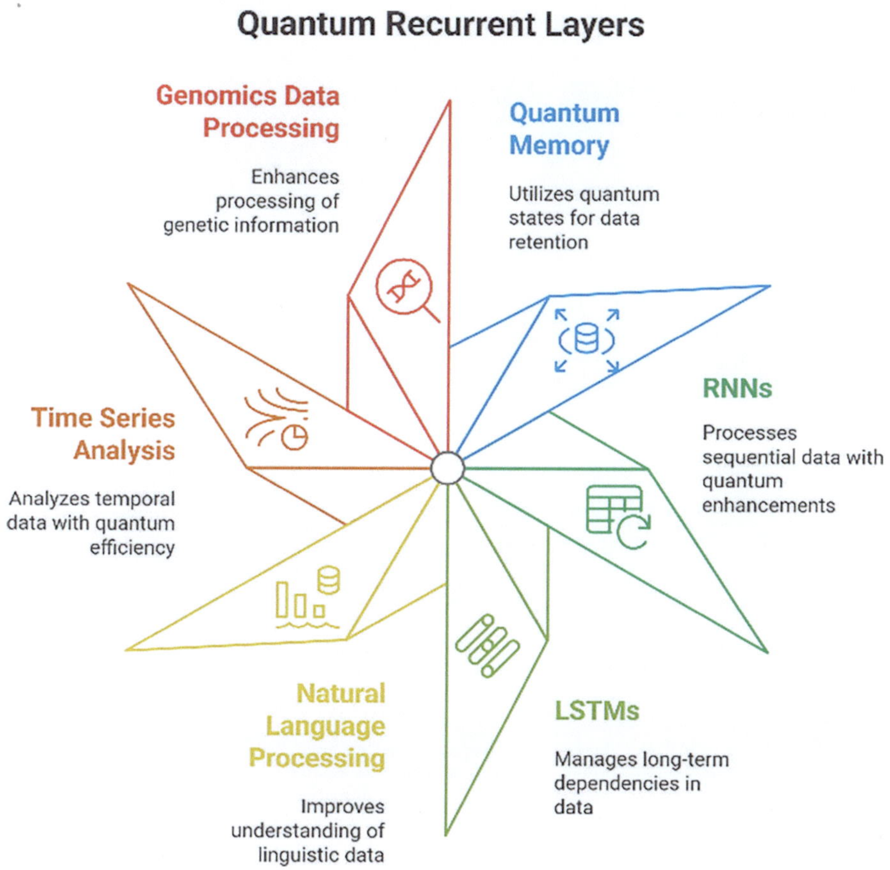

Fig. 10.15 Quantum recurrent layers

Hybrid Models combine classical and quantum layers to practically benefit from quantum speedups while retaining proven classical NN architectures. Parameters can be shared between classical and quantum parts of the network as represented in Fig. 10.17.

Specialized quantum deep learning frameworks will be needed to train these augmented models. Quantum–classical co-processors designed specifically for neural network workloads are also emerging, such as IBM's Koniku chip [12]. Overall, quantum deep learning promises to make training more powerful AI models practical by harnessing exponential quantum speedups as represented in Fig. 10.18.

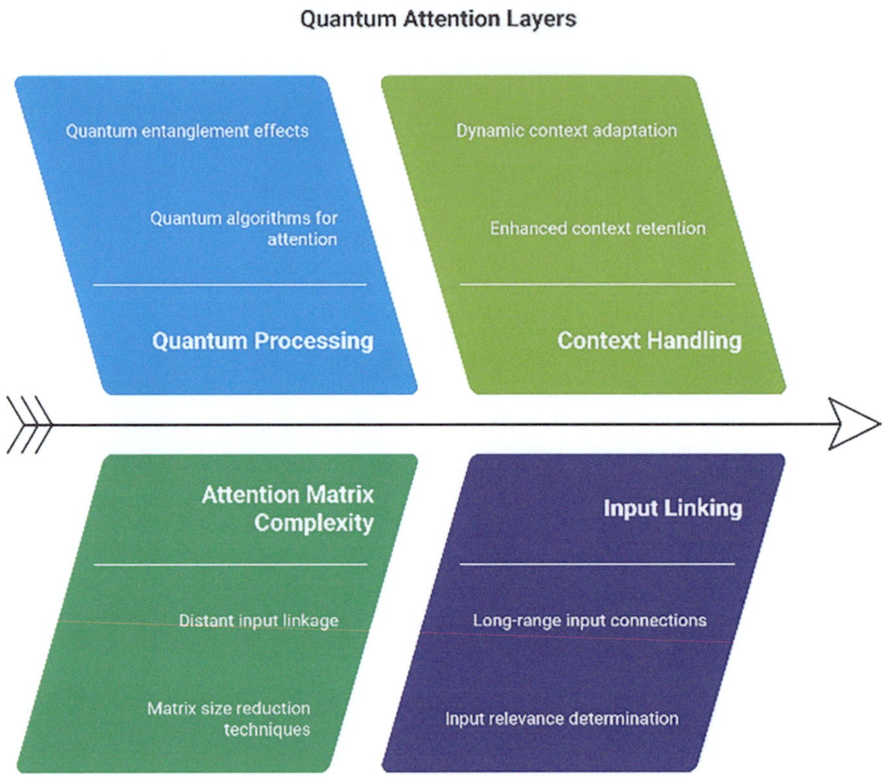

Fig. 10.16 Quantum attention layers

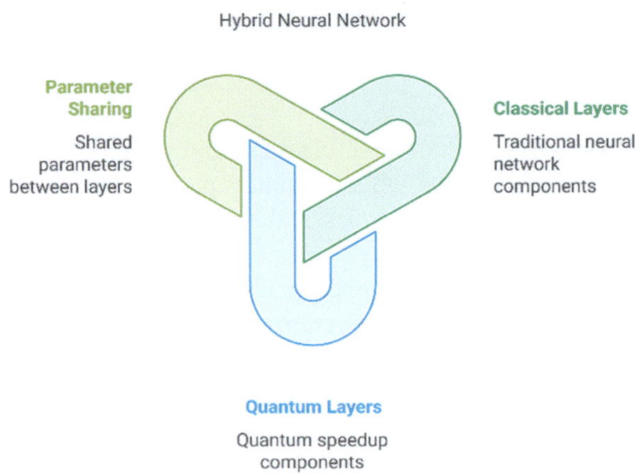

Fig. 10.17 Hybrid neural networks

Fig. 10.18 Quantum generative AI models

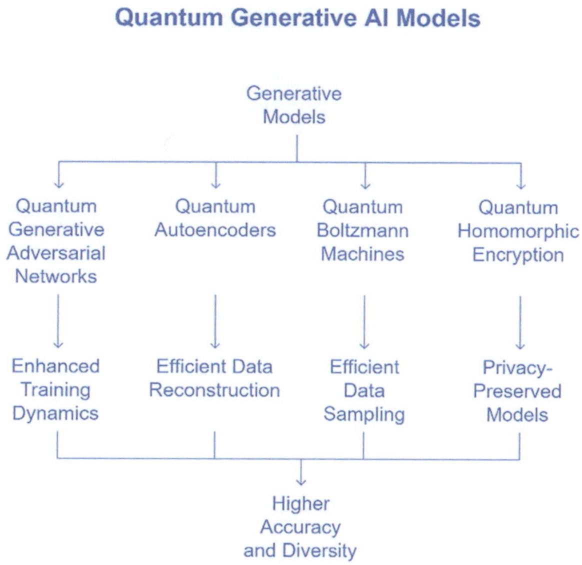

10.4 Quantum Generative AI

Generative models learn patterns from training data to generate new synthetic samples. They have applications in content creation, drug discovery, materials science, and more. Quantum computing enhances generative AI approaches [13]:

- ***Quantum Generative Adversarial Networks (QGANs)*** arm both generator and discriminator models with quantum processing capabilities. Quantum superposition enables more stable training dynamics for GANs, represented in Fig. 10.19.
- ***Quantum Autoencoders*** leverage quantum circuits to encode inputs into compressed latent representations and decode into outputs. Can reconstruct inputs even with limited encoded information (Fig. 10.20).
- ***Quantum Boltzmann Machines*** use interconnected quantum units that stochastically model high-dimensional data distributions. Allows efficiently sampling synthetic data as represented in Fig. 10.21.
- ***Quantum Homomorphic Encryption*** allows performing GAN training on encrypted data, enabling privacy-preserved generative models, as shown in Fig. 10.22.

Overall, quantum generative models can learn probabilistic representations of complex, multi-modal data distributions across exponentially large Hilbert spaces. This enables generating synthetic samples with higher accuracy and diversity compared to classical approaches.

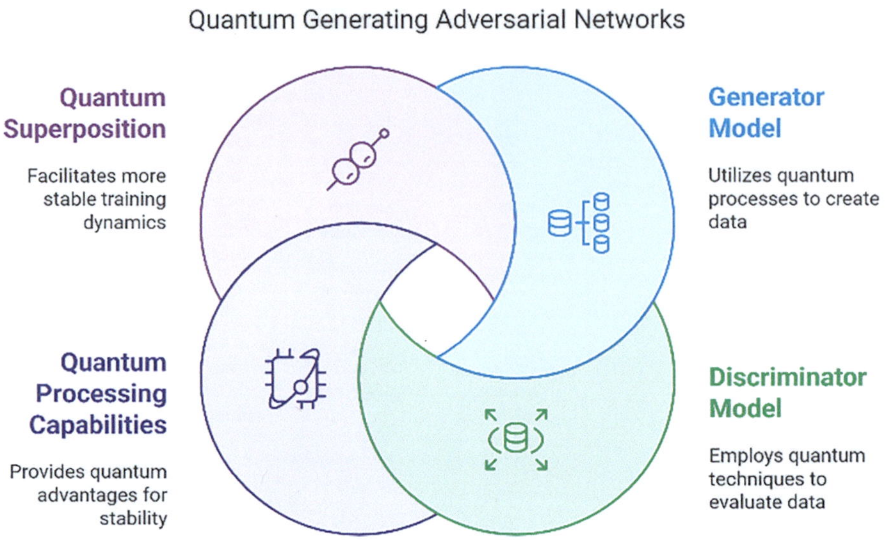

Fig. 10.19 Quantum generative adversarial networks (QGANs)

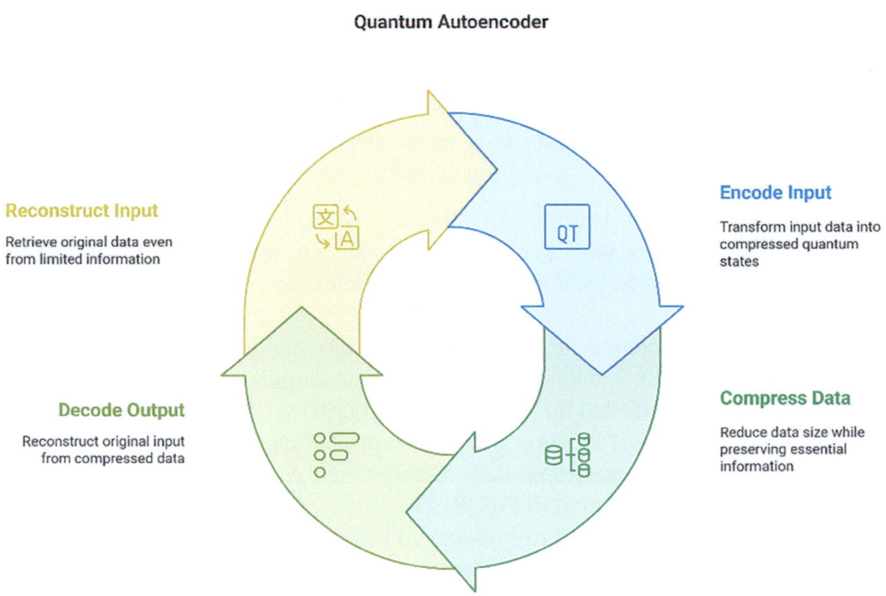

Fig. 10.20 Quantum autoencoders

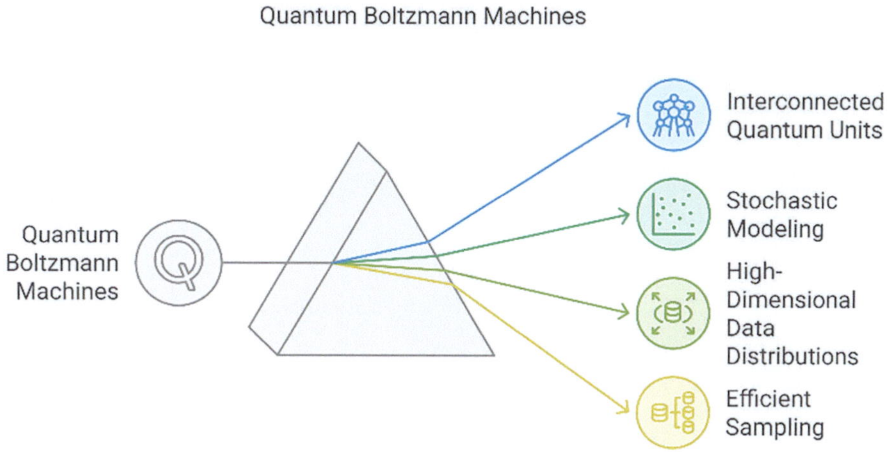

Fig. 10.21 Quantum Boltzmann machines

Fig. 10.22 Quantum homomorphic encryption

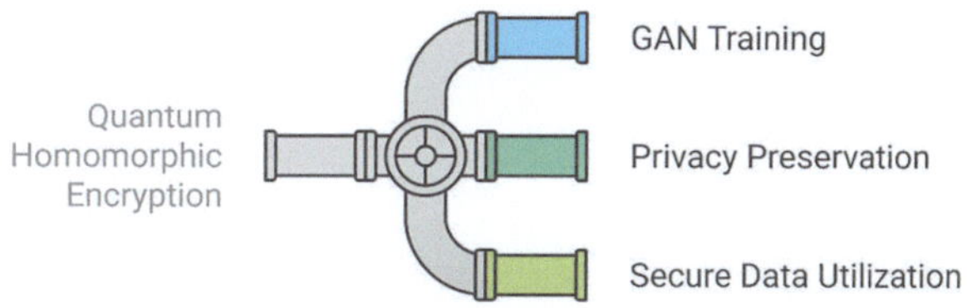

10.5 Quantum Reinforcement Learning

Reinforcement learning develops agents that optimize behaviors by interacting dynamically with environments as represented in Fig. 10.23.

Quantum computing can speed up techniques like Q-learning:

- ***Quantum Q-Learning***: Updates Q-values across multiple states and actions simultaneously via amplitude amplification, enabling exponential speedup in training agents as represented as Fig. 10.24.
- ***Quantum Policy Gradients Optimization***: Leverages quantum gradient computation across multiple trajectories to speed up optimizing policies, shown in Fig. 10.25.
- ***Quantum Value Iteration***: Uses Grover's algorithm to quadratically speed up dynamic programming in Markov decision processes as presented in Fig. 10.26.
- ***Quantum Monte Carlo***: Generates probabilistic samples of returns by quantum random walks to speedup policy evaluation as shown in Fig. 10.27.
- ***Quantum Multi-Agent RL*** Allows large populations of learning agents in superposition to solve cooperative tasks as presented in Fig. 10.28.

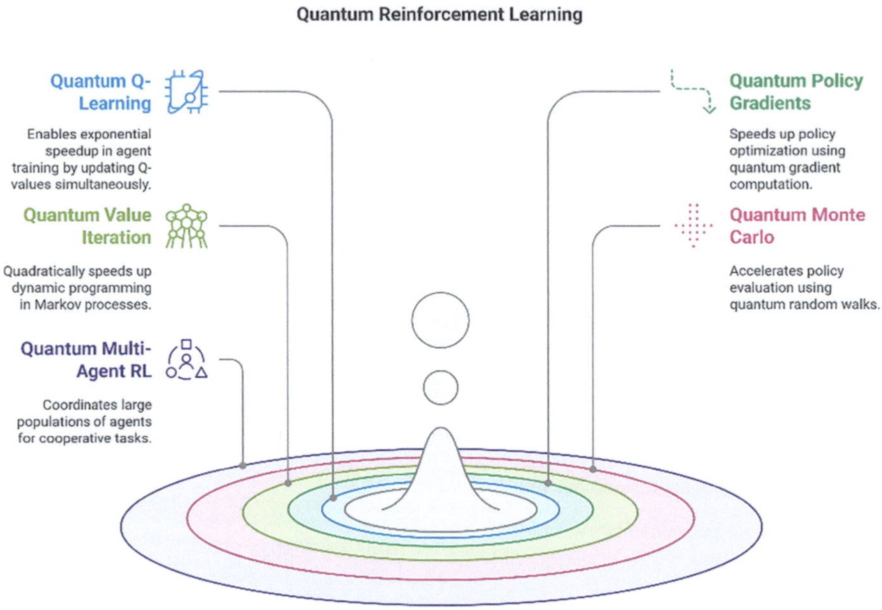

Fig. 10.23 Quantum reinforcement learning

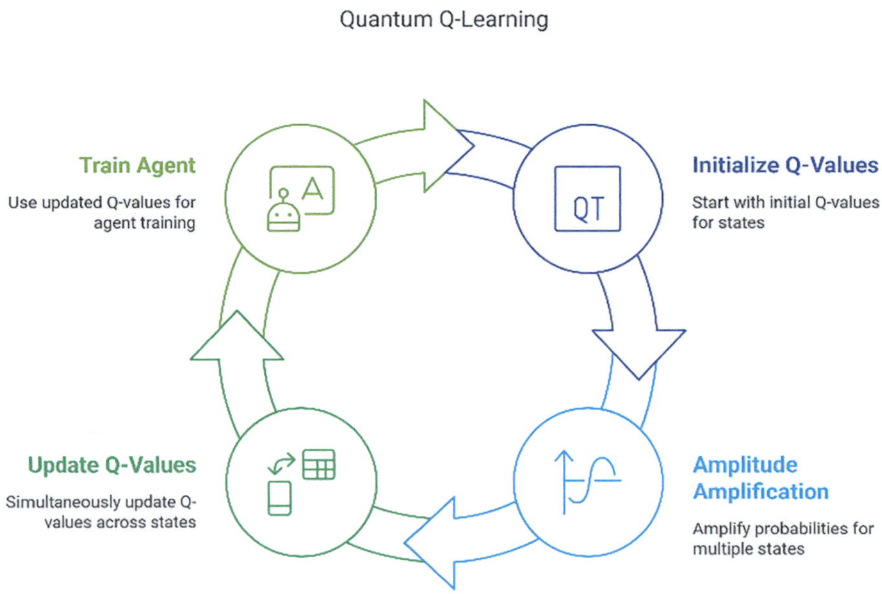

Fig. 10.24 Quantum Q-learning

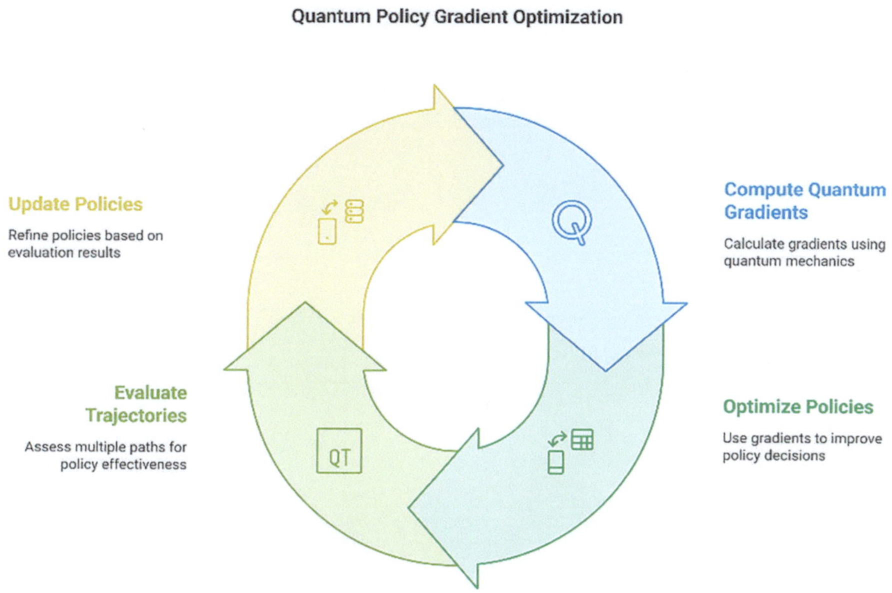

Fig. 10.25 Quantum policy gradients optimization

Quantum reinforcement learning holds promise for developing intelligent autonomous systems like robots, self-driving vehicles, drones, and manufacturing and logistics control processes [14]. Quantum agents can learn optimal policies magnitudes faster by exploiting quantum parallelism during training.

Summary

This chapter provided an overview of various promising approaches for integrating quantum computing capabilities into classical AI algorithms, models, and workflows. Quantum enhancements to AI hold the potential to unleash unprecedented predictive power and generative capabilities by combining quantum scale, speed, and dimensionality with classical machine intelligence. Realizing commercial value requires overcoming daunting challenges in building practical, scalable quantum–classical hybrid systems. Further multidisciplinary R&D across physics, computer science, and data science is vital to transform quantum AI from theory to practice. Once viable, quantum-augmented AI could become the most transformative general-purpose technology of this century.

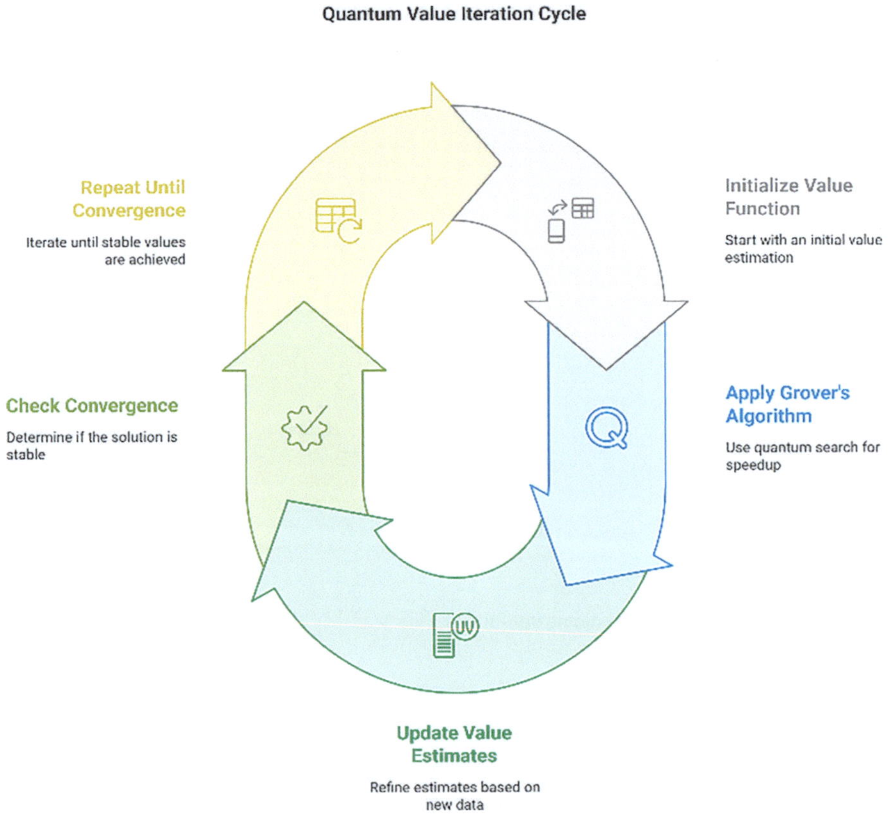

Fig. 10.26 Quantum value iteration

Fig. 10.27 Quantum Monte Carlo

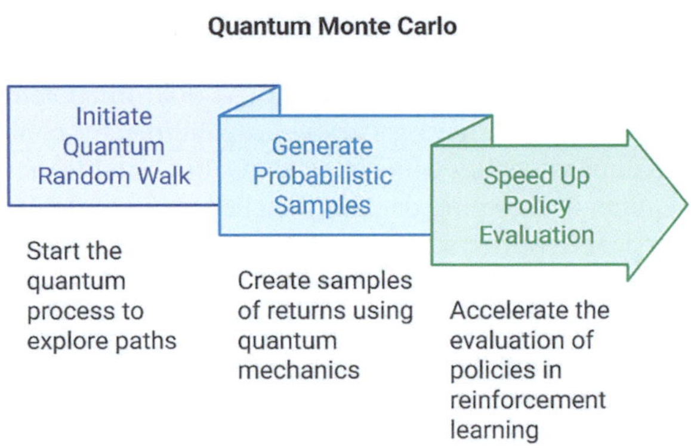

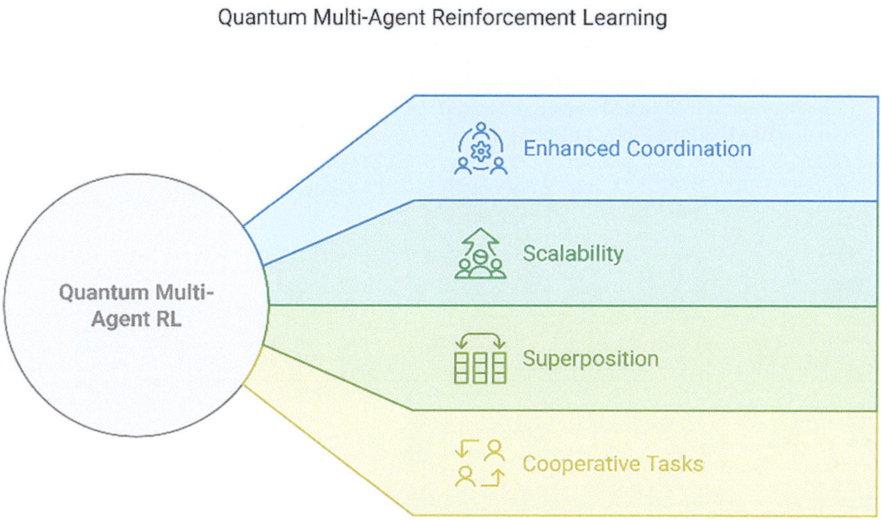

Fig. 10.28 Quantum Multi-Agent reinforcement learning

References

1. Amin, M. H., Andriyash, E., Rolfe, J., Kulchytskyy, B., & Melko, R. (2018). Quantum Boltzmann machine. *Physical Review X, 8*(2), 021050.
2. Dunjko, V., & Briegel, H. J. (2018). Machine learning & artificial intelligence in the quantum domain: A review of recent progress. *Reports on Progress in Physics, 81*(7), 074001.
3. Huggins, W., Patel, P., Whaley, K. B., & Stoudenmire, E. M. (2022). Towards quantum advantage in training deep generative models. *Quantum Science and Technology, 7*(3), 035003.
4. Meyer, D. A. (2021). Quantum algorithms for deep learning. *Philosophical Transactions of the Royal Society A, 380*(2194), 20210163.
5. Schuld, M., Bocharov, A., Svore, K. M., & Wiebe, N. (2021). Circuit-centric quantum classifiers. *Physical Review A, 103*(3), 032430.
6. Ranga, D., Rana, A., Prajapat, S., Kumar, P., Kumar, K., & Vasilakos, A. V. (2024). Quantum machine learning: Exploring the role of data encoding techniques, challenges, and future directions. *Mathematics, 12*(21), 3318.
7. Pande, M. B. (2024). A comprehensive review of data encoding techniques for quantum machine learning problems. In *2024 International conference on emerging trends in IT and engineering*.
8. Beer, K., Bondarenko, D., Farrelly, T., Osborne, T. J., Salzmann, R., Scheiermann, D., & Wolf, R. (2020). Training deep quantum neural networks. *Nature Communications, 11*, 808. https://doi.org/10.1038/s41467-020-14454-2.
9. Quiroga, D., Date, P., & Pooser, R. (2021). Discriminating quantum states with quantum machine learning. In *2021 International conference on rebooting computing*.
10. Henderson, M. P., Shakya, S., Pradhan, S., & Cook, T. (2019). Quanvolutional neural networks: Powering image recognition with quantum circuits. arXiv preprint arXiv:1904.04767. https://doi.org/10.48550/arXiv.1904.04767.
11. Chen, S. Y.-C., Yang, C.-H. H., Qi, J., Chen, P.-Y., Ma, X., & Hsieh, M.-H. (2022). Quantum long short-term memory. *IEEE Transactions on Neural Networks and Learning Systems, 33*(11), 6409–6421. https://doi.org/10.1109/TNNLS.2021.3054834.

12. Shrivastava, A., & Sharma, V. (2024). *Primer on data in quantum machine learning*—See supervised learning sub-section.
13. Lee, S., Park, D. K., Kim, J., Bang, J., & Lee, J. (2025) Quantum variational autoencoder utilizing regularized mixed-state representations. *Physical Review A, 111*(4), 042416. https://doi.org/10.1103/PhysRevA.111.042416.
14. Wang, Y., Sun, Z., Li, Z., Zhang, J., & Zhang, S. (2024) Quantum circuit autoencoder. *Physical Review A, 109*(3), 032623. https://doi.org/10.1103/PhysRevA.109.032623.

Part III
Quantum Computing Applications in Management

While the earlier parts focused on introducing quantum computing and quantum AI concepts, this part explores practical enterprise applications across management disciplines. Quantum optimization, forecasting, and simulation techniques are poised to transform everything from financial portfolio design to supply chain planning. Business leaders should actively track quantum algorithm progress, identify high impact use cases within their organizations, and pilot promising solutions on emerging quantum cloud platforms.

Quantum optimization and simulation capabilities can overcome the limitations of classical data processing, unlocking unprecedented insights and value. Managers who embrace quantum computing early can gain a sustainable competitive advantage. This section equips readers with knowledge of quantum computing's immense application potential to reimagine business processes and systems.

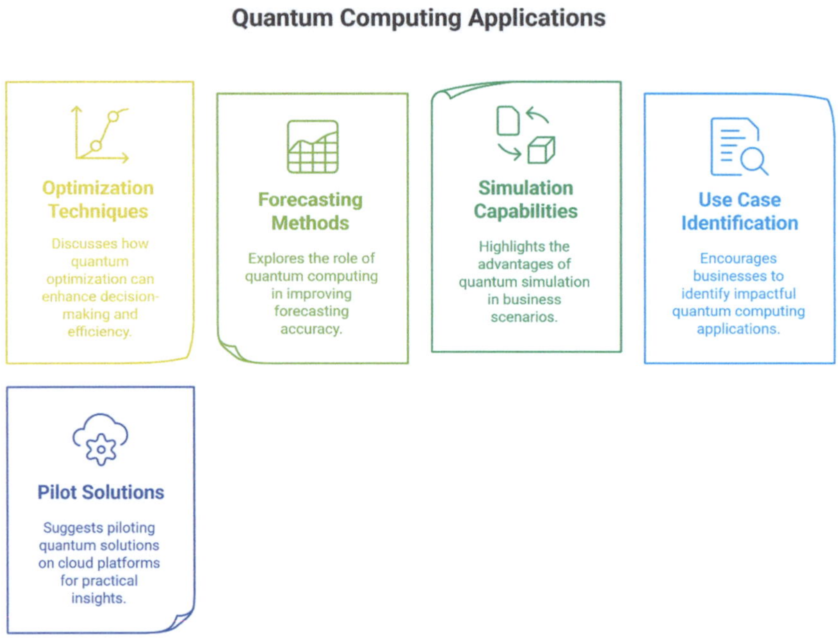

Fig. 3 Quantum computing applications

Chapter 11
Optimization and Portfolio Management

11.1 Quantum Optimization Algorithms

Many business challenges boil down to optimization problems, maximizing profit given constraints, minimizing risk given expected returns, and optimally distributing resources across alternatives. But most real-world optimization scenarios have complex search spaces with multiple local optima, making solutions hard to find and guarantee [1] (Fig. 11.1).

Quantum optimization leverages quantum mechanical effects like tunneling and superposition to traverse search spaces exponentially faster than classical methods [2]. Key quantum techniques include:

Grover's Algorithm for unsorted database search provides quadratic speedup. Useful for satisfiability problems (Fig. 11.2).

Quantum Annealing uses quantum fluctuations to tunnel through barriers in search landscapes. It is well suited for constraint satisfaction and scheduling as represented in Fig. 11.3.

The Quantum Approximate Optimization Algorithm (QAOA) cleverly mixes constraints and objectives into the energy landscape, allowing the rapid search for approximate solutions via quantum interference effects [3] as shown in Fig. 11.4.

Variational Quantum Algorithms leverage hybrid quantum–classical feedback loops to iteratively converge on optima. It allows optimization with noisy quantum hardware (Fig. 11.5).

These algorithms can find better solutions faster for optimization challenges with combinatorically large search spaces across industrial scenarios in scheduling, logistics, quantitative finance, drug discovery, materials science, and more [4].

M. Singh and B. S. Rawal, *Quantum Minds: Exploring Quantum Computing and AI Synergy*, https://doi.org/10.1007/978-981-95-2498-3_11

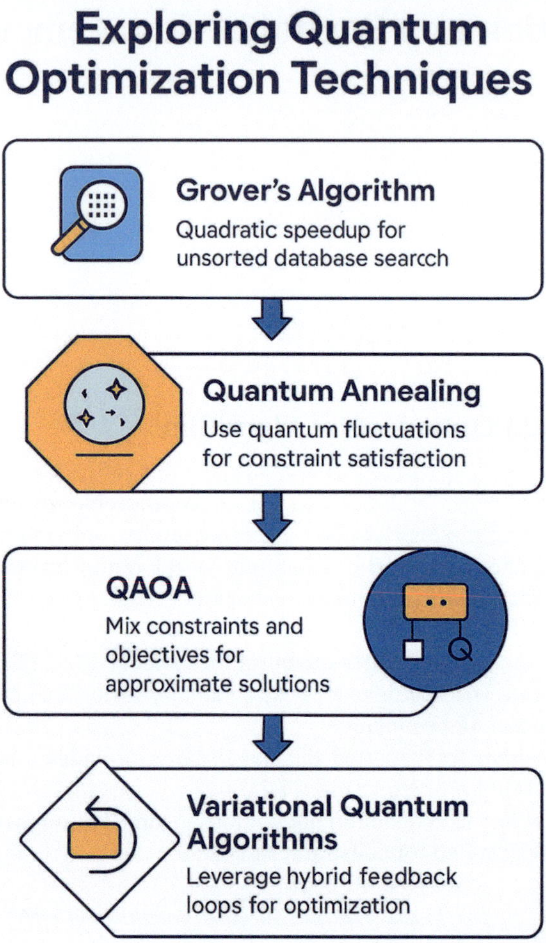

Fig. 11.1 Exploring quantum optimization techniques

11.2 Quantum Portfolio Optimization

Modern portfolio theory provides a framework for optimally investing across assets by maximizing expected returns for given risk tolerance. However, as portfolio complexity grows, finding the optimal asset allocation becomes intractable for classical solvers as represented in Fig. 11.6.

Quantum portfolio optimization extends modern portfolio theory by exploiting quantum optimization approaches like QAOA and quantum annealing to traverse

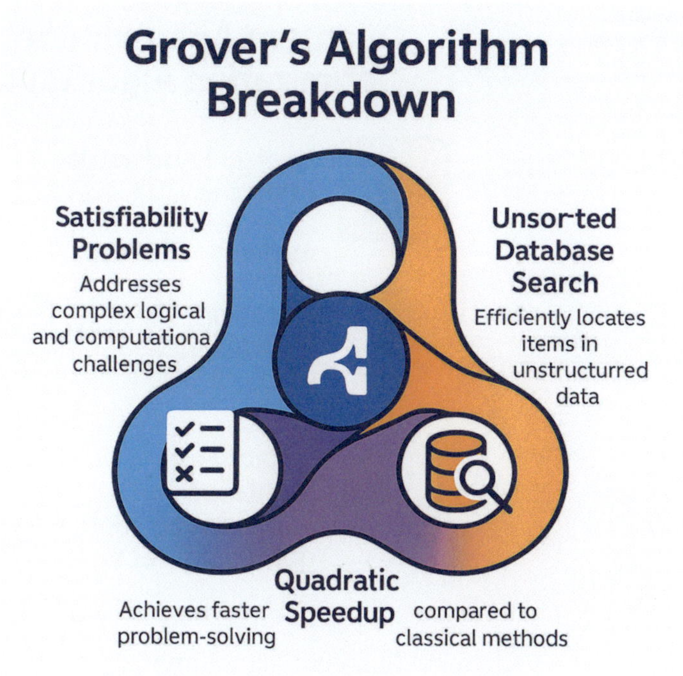

Fig. 11.2 Grover's algorithm

Fig. 11.3 Quantum annealing

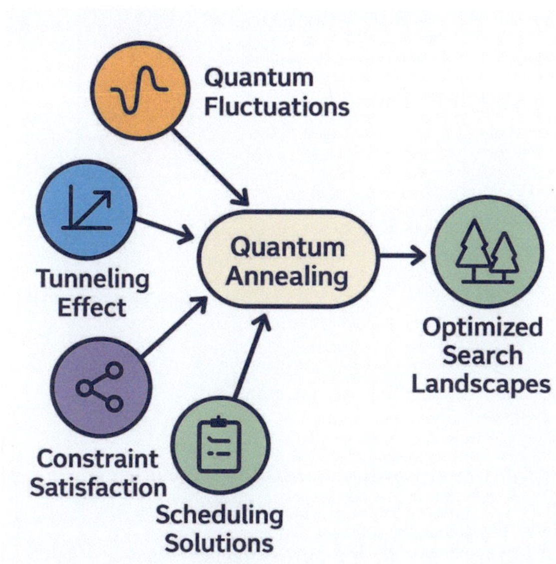

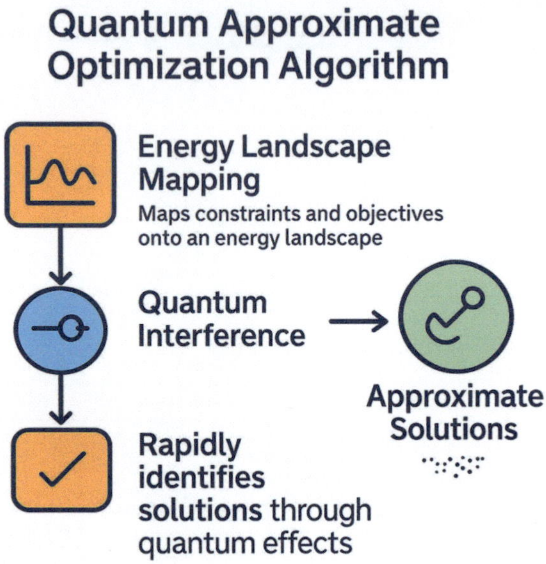

Fig. 11.4 Quantum approximate optimization algorithm

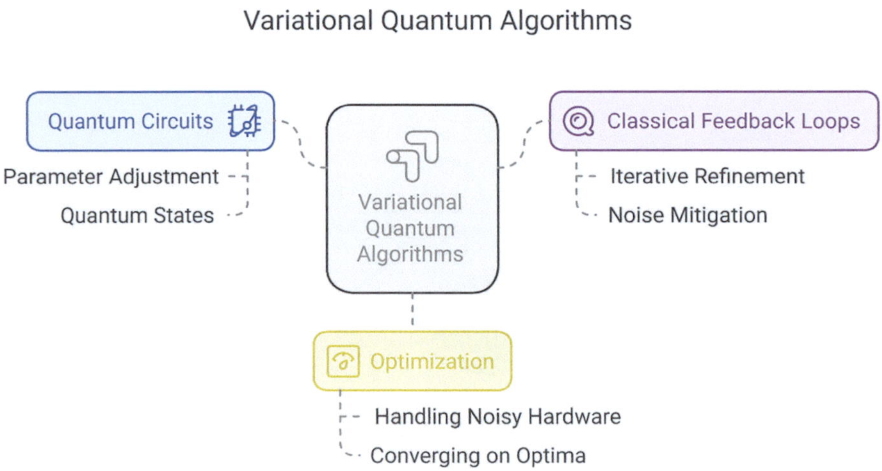

Fig. 11.5 Variational quantum annealing

portfolio superposition across the risky asset space rapidly [5]. This allows managers to:

- Optimize portfolios across larger asset universes with exponentially more asset combinations evaluated.
- Efficiently handle complex real-world constraints like transaction costs, liquidity restrictions, and diversification needs when suggesting asset allocations.

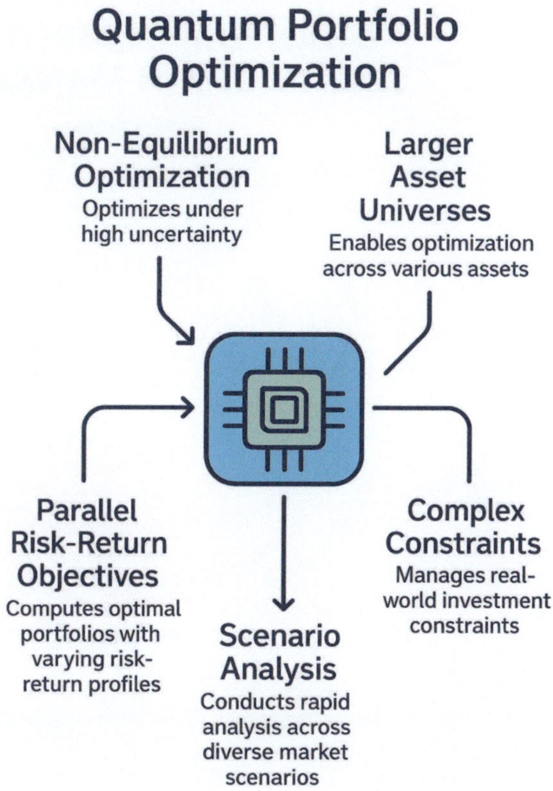

Fig. 11.6 Quantum portfolio optimization

- Compute optimal portfolios with different risk-return objectives in parallel.
- Rapid scenario analysis across simulated market environments.
- Optimize under non-equilibrium conditions and regimes with more significant uncertainty.

Quantum portfolio optimization enables investment managers to dynamically construct optimal portfolios tailored to any risk profile, constraints, and market conditions [6]. This allows them to maximize portfolio performance while adhering to client requirements.

11.3 Supply Chain Management with Quantum Computing

Global supply chains are becoming more complex, with global networks of suppliers, plants, distribution centers, and retailers interlinked through multidirectional material, information, and financial flows. Optimization and simulation techniques are vital for supply chain planning and resilience [7] (Fig. 11.7).

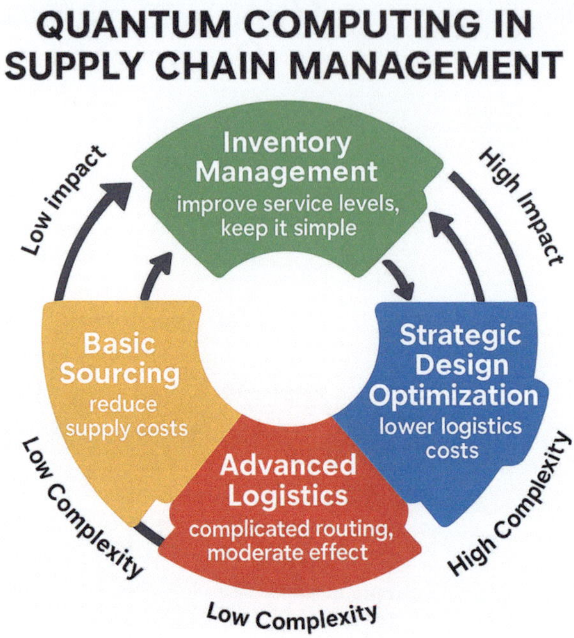

Fig. 11.7 Supply chain management with quantum computing

Quantum techniques like quantum annealing can accelerate large-scale supply chain optimization challenges like:

- *Strategic design*—optimally locating plants, distribution centers, and warehouses for minimal logistics costs and risks.
- *Sourcing*—choosing optimal suppliers globally to balance costs, quality, and reliability.
- Production planning—efficiently allocating raw materials and optimizing manufacturing schedules across facilities to meet demand.
- *Inventory management*—defining optimal order quantities, buffer stock levels across the supply network to meet service levels.
- *Logistics*—vehicle routing, fleet scheduling, and shipment planning for optimal delivery performance.

For each use case, quantum optimization provides an orders-of-magnitude speedup over classical heuristics, allowing regular supply chain re-planning as conditions change. Quantum machine learning also enables more accurate demand forecasting to improve planning [8].

Quantum computing will enable supply chain organizations to dynamically optimize their global networks in real-time as constraints and objectives evolve. This responsiveness is vital for managing supply chain disruption risks [9]. Quantum techniques provide a competitive edge in the age of highly turbulent operating environments.

References

1. Long, G. L. (2001). Grover's algorithm with zero theoretical failure rate. *Physical Review A, 64*(2), 022307. https://doi.org/10.1103/PhysRevA.64.022307.
2. Albash, T., & Lidar, D. A. (2018). Adiabatic quantum computation and quantum annealing: Theory and practice. *Reviews of Modern Physics, 90*(1), 015002. https://doi.org/10.1103/RevModPhys.90.015002.
3. Hen, I., & Spedalieri, F. M. (2016). Quantum annealing for constrained optimization. *Physical Review Applied, 5*(3), 034007. https://doi.org/10.1103/PhysRevApplied.5.034007.
4. Wurtz, J., Love, P. J., & van Dam, W. (2022). The quantum approximate optimization algorithm and the Sherrington–Kirkpatrick model at infinite size. *Quantum, 6*, 759. https://doi.org/10.22331/q-2022-07-07-759.
5. Tilly, J., Chen, H., Cao, S., Picozzi, D., Setia, K., Li, Y., Grant, E., Wossnig, L., Rungger, I., Booth, G. H., & Tennyson, J. (2022). The variational quantum eigensolver: A review of methods and practices. *Reports on Progress in Physics, 85*(7), 076001. https://doi.org/10.1038/s41598-023-45392-w.
6. Hodson, M., Morris, A., Rungger, I., & Grant, E. (2023). Best practices for portfolio optimization by quantum computing with the Variational Quantum Eigensolver. *Scientific Reports, 13*, 18502. https://doi.org/10.1038/s41598-023-45392-w
7. Orús, R., Mugel, S., & Lizaso, E. (2019). Quantum computing for finance: Overview and prospects. *Reviews in Physics, 4*, 100028. https://doi.org/10.1016/j.revip.2019.100028.
8. Sakhapov, D., & Tishaev, I. (2021). Quantum computing and supply chain optimization. *Indonesian Journal of Electrical Engineering and Informatics, 9*(3), 548–560. https://doi.org/10.11591/ijeia.v9i3.2770.
9. Macaluso, G., Gentile, C., & Venturelli, D. (2023). Quantum and quantum-inspired methods for supply chain optimization. In P. M. Pardalos & Q. Zhao (Eds.), *Machine Learning and Data Science in the Power Generation Industry* (pp. 253–280). Springer. https://doi.org/10.1007/978-3-030-90378-9_11.

Chapter 12
Risk Assessment and Fraud Detection

Quantitative risk management and fraud analytics are vital in financial services. Powerful predictive models help identify problematic transactions, minimize risks, and prevent financial crimes. Quantum techniques offer financial institutions new capabilities for risk assessment and fraud detection [1]. This section explores quantum computing applications in risk simulation, cybersecurity analytics, and anti-money laundering as represented in Fig. 12.1.

12.1 Quantum Simulation for Risk Analysis

Financial institutions like banks, insurers, and investment firms rely extensively on Monte Carlo simulations to quantify portfolio risks and stress test scenarios and estimate contingent liabilities [2]. However, running sufficiently large ensembles of simulations to capture tail risks accurately is computationally intensive, even on HPC clusters as represented in Fig. 12.2.

Quantum techniques like quantum Monte Carlo promise to speed up risk simulation. QMC leverages quantum random walks to generate probability distributions of outcomes used in scenario analysis efficiently [3]. Key techniques include:

- *Amplitude Estimation* quickly calculates the expected values of desired quantities by repeatedly applying quantum projections, reducing sampling error.
- *Quantum Generative Modeling* can efficiently produce synthetic risk parameter simulations representing stressed market conditions.
- *Quantum Conditional Value at Risk Calculation* rapidly estimates tail risk metrics like CVaR across portfolios, better optimizing risk-return tradeoffs.

With quantum accelerated Monte Carlo simulations, financial institutions can stress test their books against extreme market events across millions of simulated scenarios in minutes rather than days. This enables dynamic risk forecasts and mitigation.

M. Singh and B. S. Rawal, *Quantum Minds: Exploring Quantum Computing and AI Synergy*, https://doi.org/10.1007/978-981-95-2498-3_12

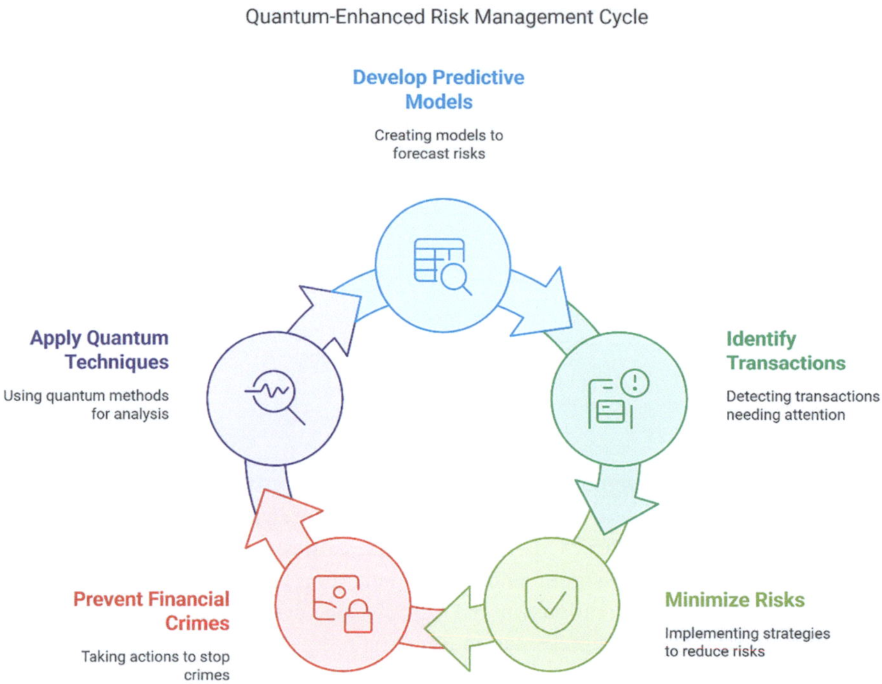

Fig. 12.1 Quantum-enhanced risk management cycle

12.2 Quantum Machine Learning for Fraud Detection

Banks lose billions of dollars yearly to payment fraud, identity theft, credit card skimming, and other financial crimes. Advanced analytics techniques like neural networks help identify suspicious patterns in massive transaction datasets as shown in Fig. 12.3.

Quantum machine learning can further boost fraud detection accuracy [4]:

- *Quantum clustering algorithms* can efficiently discern fraudulent patterns in large corpuses of queried transactions by uncovering clusters indicative of money laundering networks [5].
- *Quantum neural networks* promise faster training of deep learning models on Streaming transaction data with more predictive features learned [6].
- *Quantum graph analytics* techniques can identify anomalies in large financial networks by analyzing connections in quantum superposition. They are also better at detecting organized crime.
- *Post-quantum cryptography* will be vital for thwarting hackers with quantum computers seeking to decrypt transaction data or emails [7].

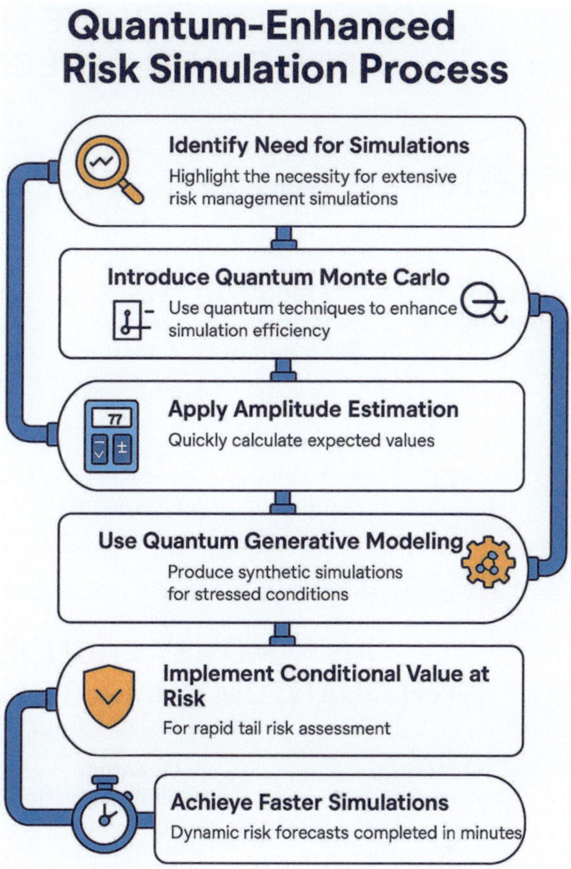

Fig. 12.2 Quantum-enhanced risk simulation process

Overall, quantum techniques provide stronger defenses against emerging threats of quantum cyberattacks while unlocking fraud insights at a scale that is not possible classically.

12.3 Anti-money Laundering Investigations

Money laundering enables trillions of dollars' worth of global financial crime annually. Investigating laundering operations requires tracing webs of "shell" companies that obscure transfers of illicit funds as depicted in Fig. 12.4. These complex corporate networks connect thousands of entities across multiple jurisdictions [8].

Quantum computing could accelerate tracing criminal money trails:

- *Quantum community detection algorithms* can efficiently uncover tightly knit groups of shell companies within broader corporate graphs [6].

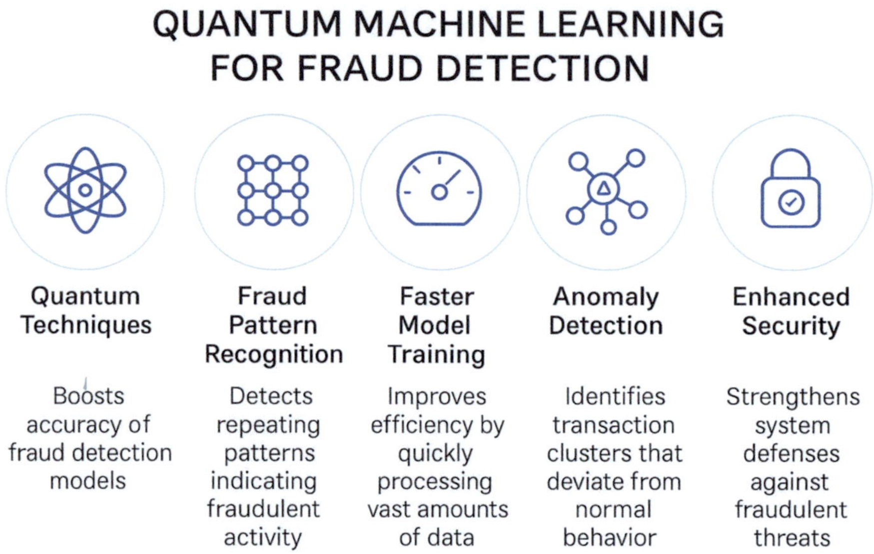

Fig. 12.3 Quantum machine learning enhances fraud detection

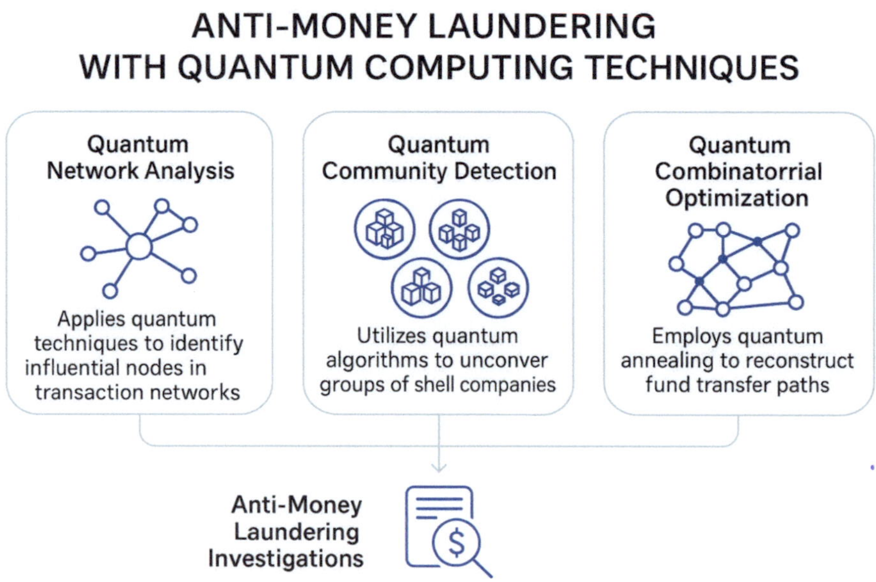

Fig. 12.4 Anti-money laundering with quantum computing

- *Quantum complex network* analysis techniques like quantum PageRank can identify highly influential nodes worth investigating in suspect transaction networks [7].
- *Quantum combinatorial optimization* approaches like quantum annealing can rapidly reconstruct likely fund transfer paths between known entities [9].

Overall, quantum network analytics applied to large corpora of financial transaction data and suspicious entity registries can greatly enhance anti-money laundering investigations and support the extraction of insights for litigation.

Summary

This section provided an overview of promising quantum computing techniques to enhance risk management and fraud detection in financial services. Quantum simulation, optimization, and machine learning solutions address key pain points around tail risk estimation, predictive fraud analytics, and tracing money laundering networks across institutions. Financial firms strategically leveraging quantum techniques early can make risk assessment more dynamic while thwarting emerging quantum cyber threats.

References

1. Stamatopoulos, N., Clader, B. D., Woerner, S., & Zeng, W. J. (2024). *Quantum risk analysis of financial derivatives*. arXiv. https://doi.org/10.48550/arXiv.2404.10088
2. Matsakos, T., & Nield, S. (2023). *Quantum Monte Carlo simulations for financial risk analytics: Scenario generation for equity, rate, and credit risk factors*. arXiv. https://doi.org/10.48550/arXiv.2303.09682
3. Adegbola, M. D., Adegbola, A. E., Amajuoyi, P., Benjamin, L. B., & Adeusi, K. B. (2024). Quantum computing and financial risk management: A theoretical review and implications. *Computer Science & IT Research Journal, 5*(6), 1210–1220.
4. Bohrium. (n.d.). *Quantum risk analysis: Efficient quantum algorithms for VaR and CVaR computations*. Bohrium.
5. Ipsos & Academia Review. (2022). Quantum computing for financial risk measurement. *Quantum Information Processing*.
6. World Economic Forum & Accenture. (2025). *Banking in the quantum era: 3 strategic shifts to watch*. World Economic Forum.
7. Moody's Insights. (2024). *Quantum computing in the financial sector: 2024 trends in review*.
8. Quantum algorithms: A new frontier in financial crime prevention (2024). arXiv.
9. BIS. (2025). *Quantum-readiness for the financial system*. BIS Papers No. 158.

Chapter 13
Decision-Making and Strategy Formulation

Making optimal decisions across complex scenarios with many alternatives and uncertain outcomes is challenging. Executives must continually formulate business strategies, balancing risk-return tradeoffs under constraints [1]. Quantum techniques offer advantages for modeling decision trees and strategizing under uncertainty (Fig. 13.1).

This section explores quantum approaches for enhancing organizational decision-making processes.

13.1 Quantum Computing for Complex Decision Trees

Decision tree analysis provides a framework for choosing between multiple options by logically mapping out consequences, payoffs, uncertainties, and evaluation criteria [2]. However, with thousands of decision nodes and branches, real-world decision trees can become huge and complex as presented in Fig. 13.2.

Classical methods struggle to traverse enormous decision spaces effectively. Quantum algorithms like quantum walks can instead efficiently construct and analyze complex decision trees by superposition across multiple paths [3, 4]:

- Faster tree traversal—Analyze decision branches in quantum parallel to identify optimal paths faster.
- Scenario simulation—Model probabilistic outcomes across decision tree branches via quantum Monte Carlo.
- Risk-return optimization—Leverage quantum optimization to identify decisions maximizing payoff under constraints and uncertainty.

This enables rational evaluation of high-impact choices across exponentially large decision spaces that are intractable for classical systems. Executives can make data-driven strategic decisions faster.

M. Singh and B. S. Rawal, *Quantum Minds: Exploring Quantum Computing and AI Synergy*, https://doi.org/10.1007/978-981-95-2498-3_13

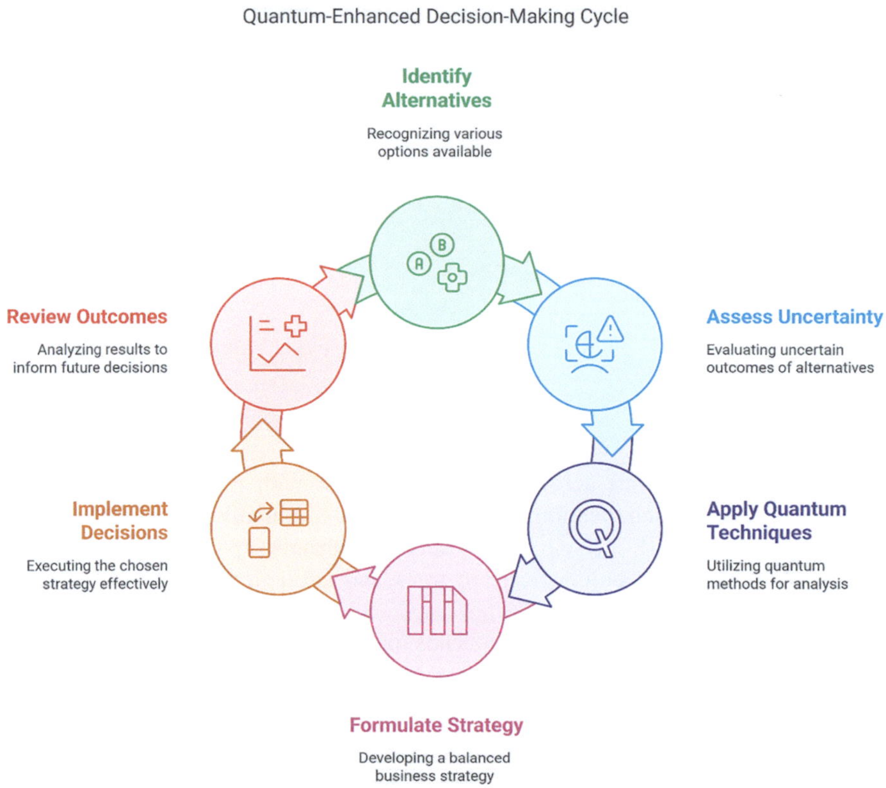

Fig. 13.1 Quantum-enhanced decision-making cycle

13.2 Quantum-Inspired Heuristics

Many business decisions rely more on executive experience and intuition than pure data-driven logic. In these scenarios, quantum principles like uncertainty and entanglement provide exciting perspectives on formulating strategy heuristically [5].

Viewing business environments as inherently uncertain rather than deterministic, as in quantum mechanics, suggests strategy should maximize flexibility and resilience [6]. Quantum entanglement also highlights interdependencies between strategic choices, competitors, and external factors that produce emergent effects as represented in Fig. 13.3.

Quantum-inspired business strategy emphasizes adaptiveness, agility, optionality, and ecosystem perspective. While qualitative, quantum analogies can enhance strategic thinking beyond simplistic linear cause-and-effect assumptions.

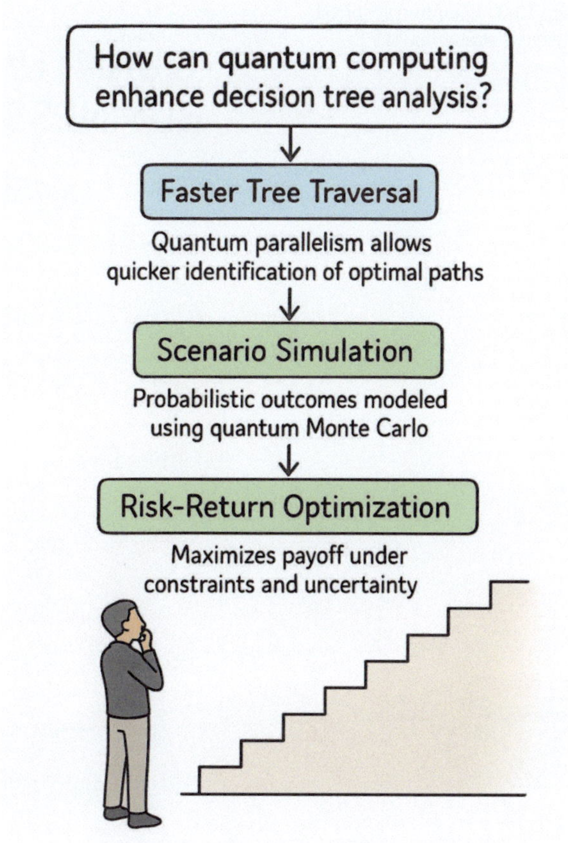

Fig. 13.2 Quantum computing enhance decision tree

13.3 Quantum Algorithms for Strategic Planning

Devising business plans involves optimizing resource allocation across strategic initiatives to maximize value. This requires reconciling business objectives, forecast uncertainties, budgets, and operational constraints [7]. As strategy scenarios grow more complex, identifying optimal activity portfolios becomes intractable as Fig. 13.4.

Quantum computing could enhance strategic planning processes (Fig. 13.5):

- *Quantum Monte Carlo* can rapidly stress test prospective strategies across millions of market scenarios to estimate expected performance [8].
- *Quantum optimization* approaches can efficiently allocate resources across thousands of potential strategic initiatives to identify plans maximizing objectives [9].
- *Quantum machine learning* applied to past results can better predict the performance of new strategic options [10].

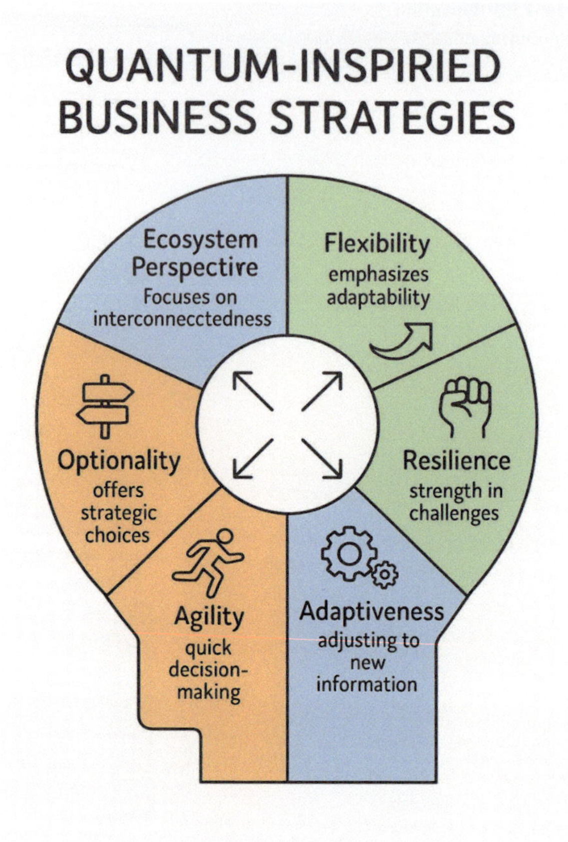

Fig. 13.3 Quantum-inspired business strategies

- *Quantum simulation* of competitor reactions can stress test for robustness [11].

With quantum systems, strategy teams can model exponentially more alternatives and scenarios when formulating business plans. This enables the devising of robust strategies aligned to corporate goals despite market uncertainties.

Summary

This section provided an overview of how quantum techniques can enhance organizational decision-making and strategy formulation. Quantum computing allows efficiently tackling complex decision problems and strategic optimizations that strain classical systems. Business leaders should look for applications of quantum decision theory and quantum-inspired strategy approaches.

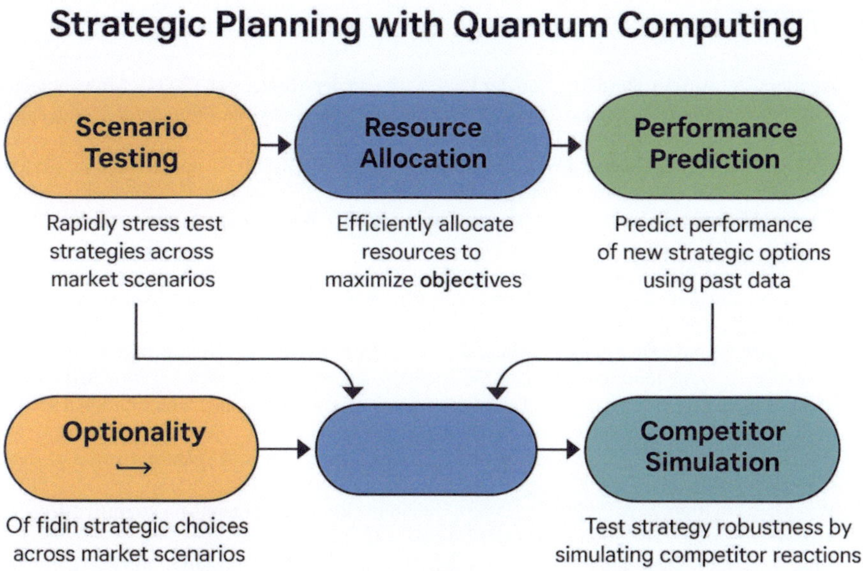

Fig. 13.4 Strategic planning with quantum computing

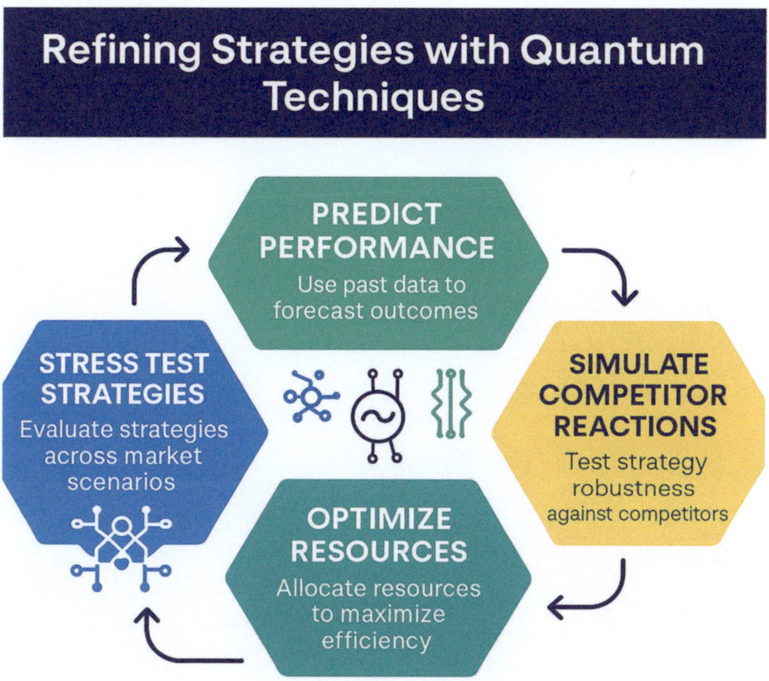

Fig. 13.5 Refining strategies with quantum techniques

References

1. Aguilera, A., van der Starre, R., van der Poel, J., & van der Meulen, M. (2024). Multi-objective portfolio optimization using a quantum annealer. *Journal of Risk and Financial Management, 17*(4), 215. https://doi.org/10.3390/jrfm17040215
2. Roffe, J. (2018). Quantum physics, strategy and POSSIBILITIES: potentiality in strategic management. *Strategic Management Review, 1*(1).
3. Thicke, M. (2021). Using decision trees as a strategy formulation framework. *Journal of Strategy and Management*.
4. Deutsch, D. (1999). Quantum theory of probability and decisions. *Proceedings of the Royal Society of London. Series A: Mathematical, Physical and Engineering Sciences, 455*(1988), 3129–3137.
5. Narayanan, A., & Menon, V. (1998, July). Quantum-inspired creative problem solving. In *Proceedings of the annual meeting of the cognitive science society* (Vol. 20, No. 20).
6. Menon, V., & Dagli, C. H. (2017). Leveraging quantum computing to enhance national security decision making. *IEEE Security & Privacy, 15*(5), 33–41.
7. Xin, L., & Xin, H. (2022). *Decision-making under uncertainty: A quantum value operator approach*. arXiv.
8. Widdows, D., Rani, J., & Pothos, E. (2023). Quantum circuit components for cognitive decision-making. *Entropy, 25*(4), 548. arXivMDPI+1arXiv+1
9. Klug, F. (2023). *Quantum optimization algorithms in operations research: Methods, applications, and implications*. arXiv
10. Morales, M. E. S., Biamonte, J. D., & Zimborás, Z. (2019). On the universality of the quantum approximate optimization algorithm. *Quantum Information Processing*. Wikipedia
11. Blekos, K., et al. (2024). *A review on quantum approximate optimization algorithm and its variants*. Physics Reports.

Part IV
Quantum AI Implementation

The previous parts introduced quantum computing concepts, quantum machine learning techniques, and enterprise applications for management professionals. However, translating this potential into practical business advantage requires usable quantum hardware and software platforms. This concluding part provides an overview of the quantum technology landscape emerging today in areas like quantum hardware, cloud services and software stacks.

Understanding the capabilities and limitations of current quantum computers will help managers set realistic expectations on harnessing quantum techniques for business impact. Mapping quantum solution vendors to potential use cases is also vital for pilot projects. Overall, this section equips readers with knowledge to navigate the complex, rapidly evolving quantum ecosystem.

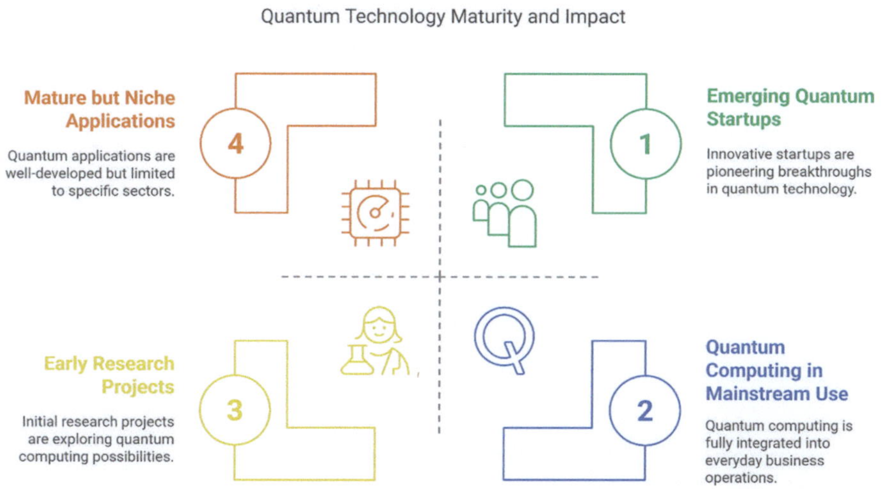

Chapter 14
Quantum Hardware Landscape

Practical quantum computing requires engineered hardware able to manipulate quantum information encoded in physical systems like photons, electrons, or atoms. Building this functionality involves integrating components like qubits, controllers, amplifiers, modulators, cryogenics, and interconnects as represented in Fig. 14.1. Multiple qubit modalities and architectural approaches are being researched. This chapter discussed prominent quantum computing platforms today and compares their capabilities [1].

14.1 Types of Quantum Processors

Multiple quantum processors are using in quantum computing such as represented in Fig. 14.2.

- *Superconducting Qubits* use Josephson junction circuits cooled to near absolute zero temperature. They were pioneered by companies like IBM, Google, and Rigetti. Benefits include easier manufacturability at scale, but qubits have higher error rates [2].
- *Trapped Ion Qubits* utilize individual charged atoms suspended by electric fields and manipulated with lasers. They are led by IonQ and offer lower error rates, but scaling interconnected qubits is challenging [3].
- *Photonics Qubits* encode quantum states in properties of photons, like polarization. They allow transmitting quantum information over long distances for communication. Qubit states are easier to produce but hard to store [3].
- *Topological Qubits* based on exotic quasiparticles and their statistical interactions offer more inherent error protection. They are still in the early stage of research, pioneered by Microsoft, and extremely challenging to engineer [4].

M. Singh and B. S. Rawal, *Quantum Minds: Exploring Quantum Computing and AI Synergy*, https://doi.org/10.1007/978-981-95-2498-3_14

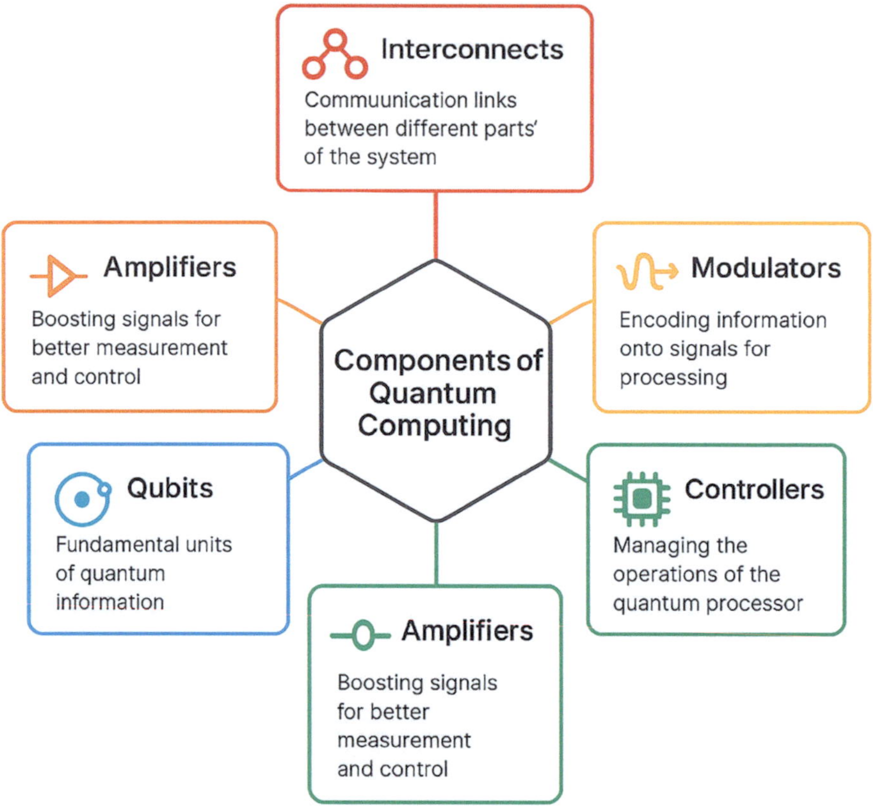

Fig. 14.1 Basic components of quantum computing

- *Quantum Annealers*, like D-Wave systems, encode optimization problems in quantum tunneling effects. They act as special-purpose quantum processing units but have limited general computing capabilities [5].

Each approach has unique strengths and weaknesses. The optimal qubit modality depends on specific use cases, superconducting systems currently dominate broad-application quantum computing, while quantum annealers and photonic chips target special functions.

14.2 Quantum Volume Metric

The quantum volume metric provides a hardware-agnostic measure of a quantum system's computational power based on factors like [6]:

- Number of qubits

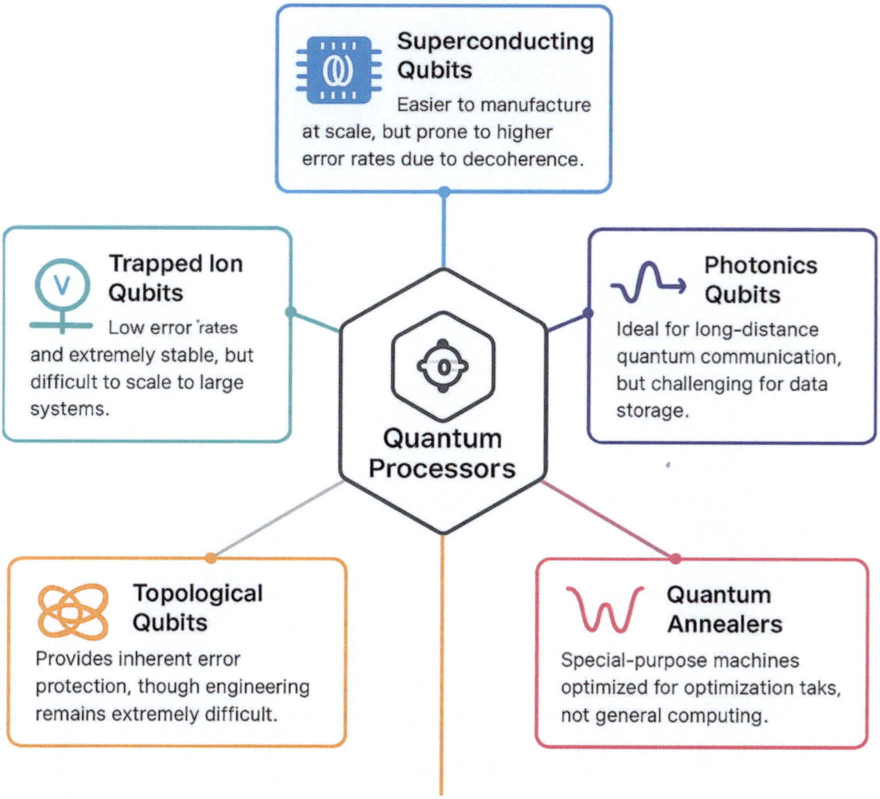

Fig. 14.2 Type of quantum processors

- Low error rates per operation
- Gate fidelity and flexibility
- Crosstalk and connectivity between qubits
- Circuit depth for the complexity of algorithms executed.

A higher quantum volume correlates with the ability to run advanced quantum circuits and workloads as shown in Fig. 14.3.

Superconducting quantum computers have the highest demonstrated quantum volume, but alternative qubit modalities continue maturing rapidly.

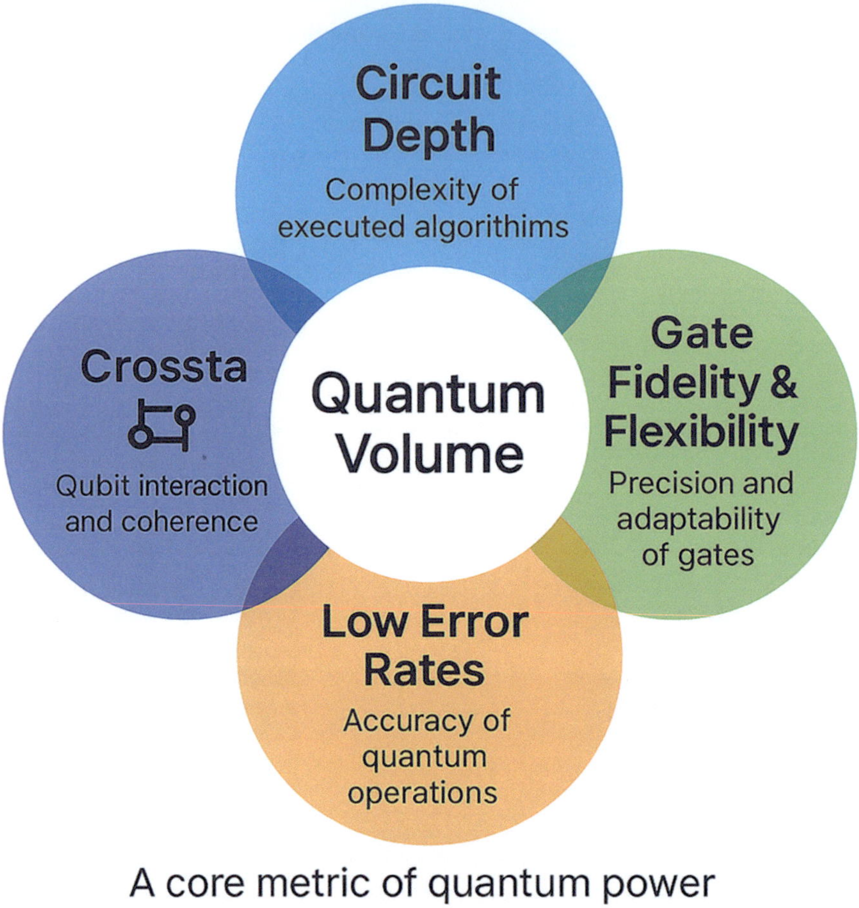

Fig. 14.3 Quantum volume and it's factors

14.3 Quantum Error Correction

Qubits are incredibly fragile, and ambient noise causes errors to accumulate rapidly during quantum algorithms. This leads to incorrect output unless errors are actively detected and corrected. Quantum error correction (QEC) techniques are thus essential for practical large-scale quantum computing.

QEC encodes logical qubit states distributed across multiple physical qubits. Even if some physical qubits fail, the encoded state can be recovered. This introduces redundancy to tolerate errors, like classical error correction in digital circuits and communication. However, complex QEC architectures with many auxiliary qubits are needed for meaningful error suppression. Quantum volume metrics rising above 4096 and beyond will likely require QEC [7, 8] (Fig. 14.4).

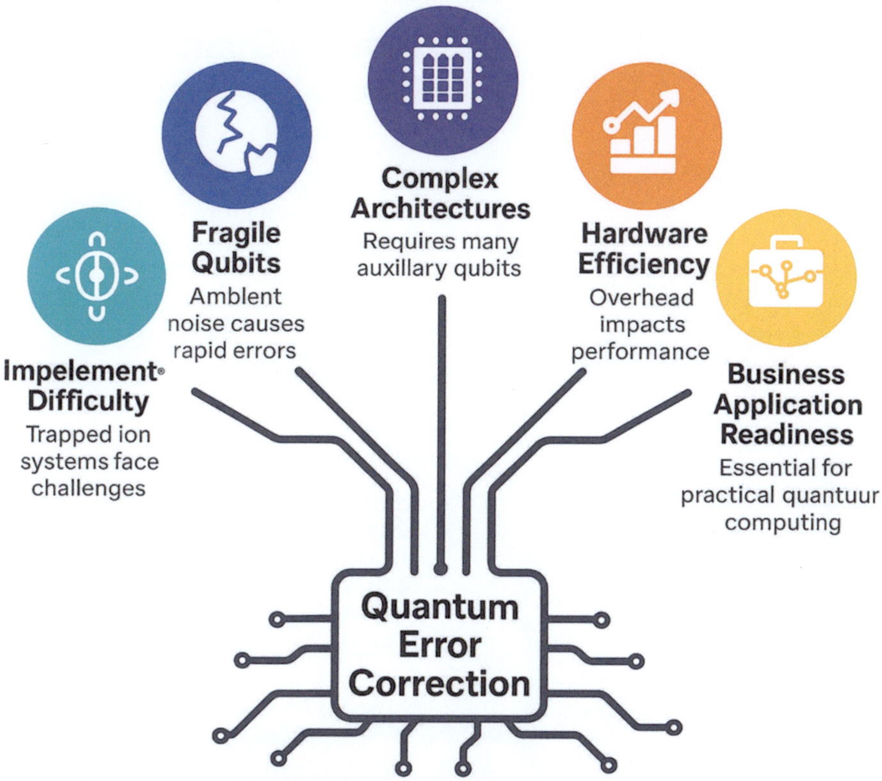

Fig. 14.4 Quantum error correction

The overhead of QEC mechanisms is a crucial hardware efficiency factor. Superconducting platforms have higher physical qubit overheads but benefit from more flexible encode/decode implementations. In theory, trapped ion systems need lower physical qubit overheads, but implementing QEC sequences is challenging. Overall, advancing QEC capabilities will be critical to harnessing quantum computing power for business applications.

14.4 Quantum Cloud Platforms

Developing in-house quantum infrastructure is still prohibitively complex and expensive for most enterprises. Managed quantum cloud services offer a more accessible pathway to explore quantum computing [9].

Figure 14.5 has shown the leading providers which include as follows:

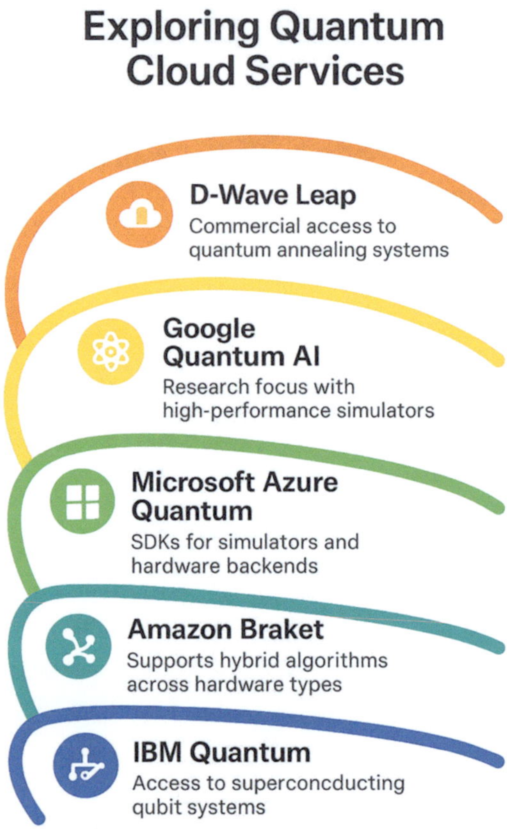

Fig. 14.5 Quantum cloud services

- *IBM Quantum* is the most mature quantum cloud platform, with over 20 quantum systems accessible via a virtual programming lab. It uses superconducting qubit systems.
- *Amazon Braket* allows testing hybrid algorithms across multiple quantum hardware types, including systems by IonQ, Rigetti, and D-Wave.
- *Microsoft Azure Quantum*—Provides SDKs for developing quantum programs on various simulators and hardware backends.
- *Google Quantum AI*—Research-focused cloud offering high-performance simulators for algorithm testing. Integrates with other Google Cloud services.
- *D-Wave Leap*—Commercial access to D-Wave's quantum annealing systems for optimization workloads.

These cloud services lower barriers to starting quantum software development and testing use cases on real quantum processors. High-level SDKs and simulators abstract away low-level hardware details [10]. Integrating quantum workflows with

Fig. 14.6 Quantum computing cloud services analysis

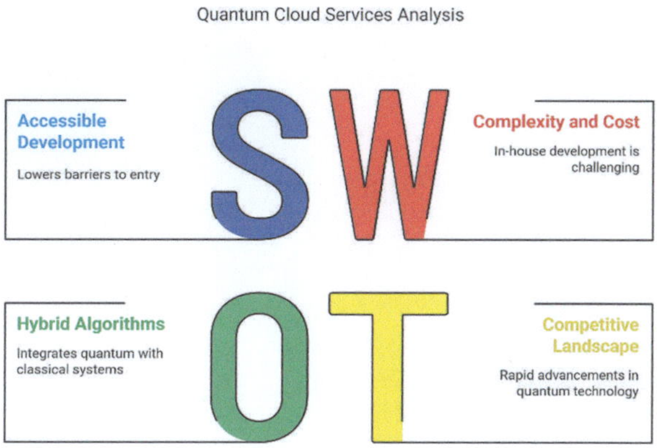

classical cloud infrastructure enables viable hybrid algorithms as represented in Fig. 14.6.

14.5 Quantum Software Stacks

Specialized software toolkits and languages are required to develop and execute quantum programs as shown in Fig. 14.7.

Here are some popular options include:

- *Qiskit*—IBM's open-source quantum programming framework with Python SDK.
- *Cirq*—Python library for writing quantum programs from Google AI Quantum.
- *Amazon Braket SDK*—Managed AWS SDK for quantum development.
- *Microsoft QDK*—Quantum Development Kit for programming on Azure Quantum.
- *D-Wave Ocean*—Full stack software suite for the D-Wave quantum annealer systems.
- *Rigetti Grove/PyQuil*—Toolkits for coding in Python and running on Rigetti's Quantum Cloud Services.

These stacks help developers construct and manipulate quantum circuits, optimize mappings to backends, simulate algorithm execution, interface with quantum processors via cloud services, and analyze outputs. Higher-level abstractions and languages will emerge for easier application development [11].

Summary

This concluding section surveyed enabling hardware, software, and cloud platforms vital for unlocking practical business advantages from quantum techniques. While quantum computing is still maturing, rapid advances bring powerful capabilities within reach. Managers should actively track technical progress, especially in areas

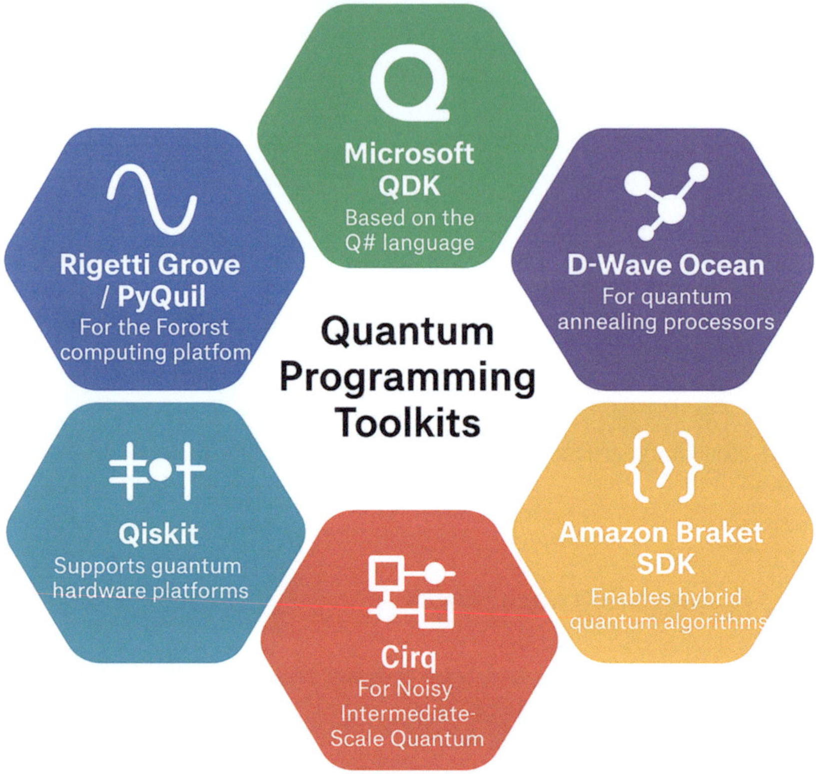

Fig. 14.7 Quantum computing programming toolkits

like quantum volume, error correction, and cloud integration that will drive real-world impact. Identifying high-value applications early combined with pilots using quantum cloud services will help chart a path to quantum advantage.

References

1. Alexeev, Y., Bacon, D., Brown, K. R., Calderbank, R., Carr, L. D., Chong, F. T., DeMarco, B., Englund, D., Farhi, E., Fefferman, B., & Gorshkov, A. V. (2021). Quantum computer systems for scientific discovery. *PRX Quantum, 2*(1), 017001.
2. Bharti, K., Cervera-Lierta, A., Kyaw, T. H., Haug, T., Alperin-Lea, S., Anand, A., Degroote, M., Heimonen, H., Kottmann, J. S., Menke, T., & Mok, W. K. (2022). Noisy intermediate-scale quantum algorithms. *Reviews of Modern Physics, 94*(1), 015004.
3. LaRose, R. (2019). Overview and comparison of gate level quantum software platforms. *Quantum, 3*, 130.
4. Preskill, J. (2018). Quantum computing in the NISQ era and beyond. *Quantum, 2*, 79.

5. Svore, K. M., Geller, M., Troyer, M., Azariah, J., Granade, C., Heim, B., Wiebe, N., et al. (2022). Qubits for business: The opportunities and challenges that quantum computing presents for businesses today. *ACM Transactions on Quantum Computing, 3*(1), 1–34.
6. Miller, K., Broomfield, C., Cox, A., Kinast, J., & Rodenburg, B. (2022). An improved volumetric metric for quantum computers via more representative quantum circuit shapes. arXiv: 2207.02315.
7. Moll, N., Barkoutsos, P., Bishop, L. S., Chow, J. M., & Cross, A. (2018). Quantum optimization using variational algorithms on near-term quantum devices. *Quantum Science and Technology*.
8. Google's Willow team. (2024, Feb). Quantum error correction below the surface code threshold. *Nature, 62* (Surface code milestone).
9. Cai, Z., Babbush, R., Benjamin, S. C., Endo, S., Huggins, W. J., Li, Y., McClean, J. R., & O'Brien, T. E. (2022). *Quantum Error Mitigation*. arXiv:2210.00921.
10. Wang, Z., & Tang, H. (2024). Artificial intelligence for quantum error correction: a comprehensive review. arXiv:2412.20380.
11. Nguyen, H. T., Krishnan, P., Krishnaswamy, D., Usman, M., & Buyya, R. (2024). Quantum cloud computing: a review, open problems, and future directions. *ACM Computing Surveys*.

Chapter 15
Programming Quantum Computers

Quantum computing promises immense processing capabilities beyond classical systems. However, harnessing this power requires mastering the art and science of programming quantum computers using specialized languages, simulators, and hardware interfaces. This concluding chapter provides a comprehensive overview of quantum programming concepts, development workflows, and debugging techniques for software engineers and data scientists as represented in Fig. 15.1 [1, 2].

This chapter provides foundational literacy for beginning practical quantum software development using frameworks like Qiskit, Cirq, or Amazon Braket [3]. They will also discuss methodologies and best practices for debugging quantum code, optimizing quantum circuits, and integrating with classical programs. This chapter equips technical professionals with key knowledge and skills to productively leverage quantum computing to deliver business impact [4].

15.1 Quantum Programming Languages and Frameworks

Classical computers execute imperative programs that specify sequential instructions to manipulate stored data. In contrast, quantum programs define circuit-based transformations on quantum states, harnessing effects like superposition, entanglement, and interference [5]. This dichotomy necessitates new quantum-specific programming paradigms (Fig. 15.2) [6].

Low-level quantum languages provide primitives to describe sequences of quantum logic gates operating on qubits. High-level languages build abstractions for modules, functions, and memory on top of these primitives. Hybrid classical-quantum languages also integrate quantum subroutines within conventional code [7]. Figure 15.3 has shown the Prominent quantum programming languages and frameworks as follows [8]:

M. Singh and B. S. Rawal, *Quantum Minds: Exploring Quantum Computing and AI Synergy*, https://doi.org/10.1007/978-981-95-2498-3_15

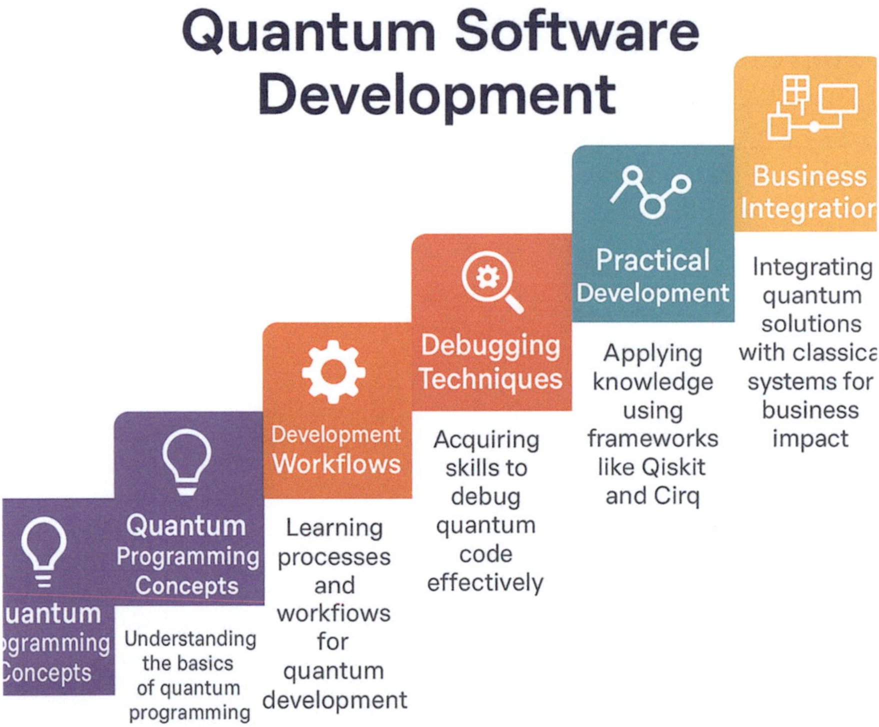

Fig. 15.1 Quantum computing software development

- ***Qiskit***—Open-source Python framework with rich libraries for composing and manipulating quantum circuits.
- ***Cirq***—Python libraries from Google for building and optimizing quantum programs.
- ***Amazon Braket***—Fully managed AWS SDK for quantum development.
- ***Microsoft Q#***—Domain-specific language with classical host language bindings.
- ***D-Wave Ocean***—Full stack software suite for the D-Wave quantum annealer.
- ***Rigetti Grove/PyQuil***—Toolkits for coding in Python and running on Rigetti quantum computers.
- ***QCL (Quantum Computation Language)***—High-level functional quantum programming language from Cambridge Quantum.
- ***Quipper***—Embedded functional language for quantum circuits with classical host language integration.

These frameworks provide modular libraries of quantum gates, optimizers, simulators, and cloud backends to construct, refine, debug, and run quantum algorithms. Higher-level quantum programming languages will continue evolving productivity abstractions like automatic circuit synthesis and error correction [9, 10].

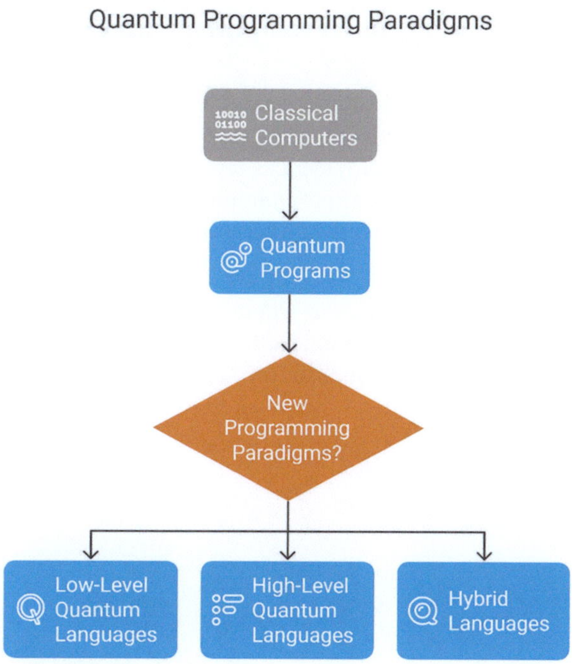

Fig. 15.2 Quantum computing programming paradigms

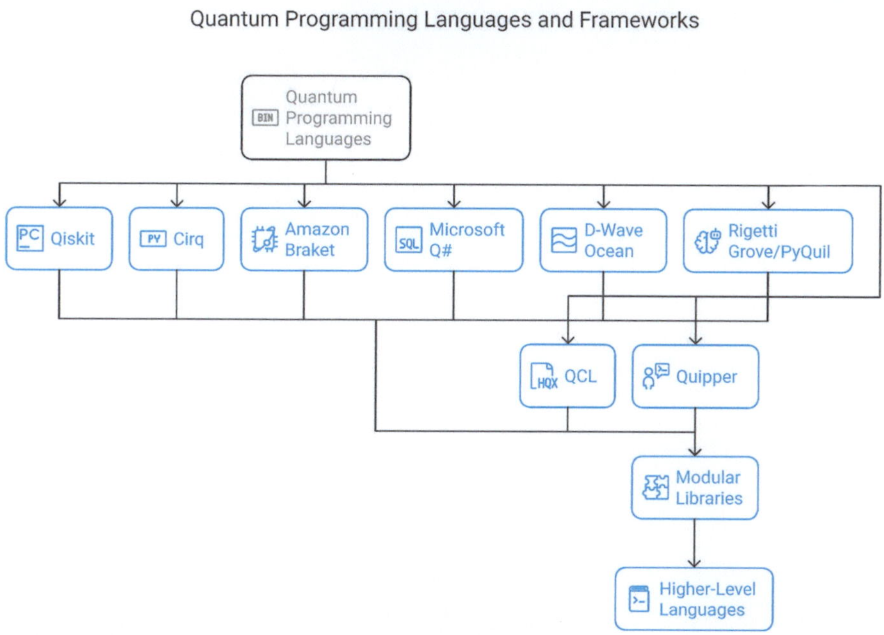

Fig. 15.3 Quantum computing programming languages and frameworks

15.2 Quantum Development Lifecycle

The quantum application development lifecycle shares similarities with classical software—requirements analysis, high-level design, detailed implementation, testing, debugging, and deployment as represented in Fig. 15.4 [8].

However, quantum programming demands specialized workflows, given unique constraints around hardware access, state fragility, and measurement [11]. Key phases include:

- *Problem Formulation*—Identify computational bottlenecks in targeted business use cases that quantum techniques can potentially accelerate—frame problems in applicable mathematical formalisms.
- *Algorithm Design*—Based on mathematical problem representation, design candidate quantum circuits, and sequences to implement desired state transformations and computations.

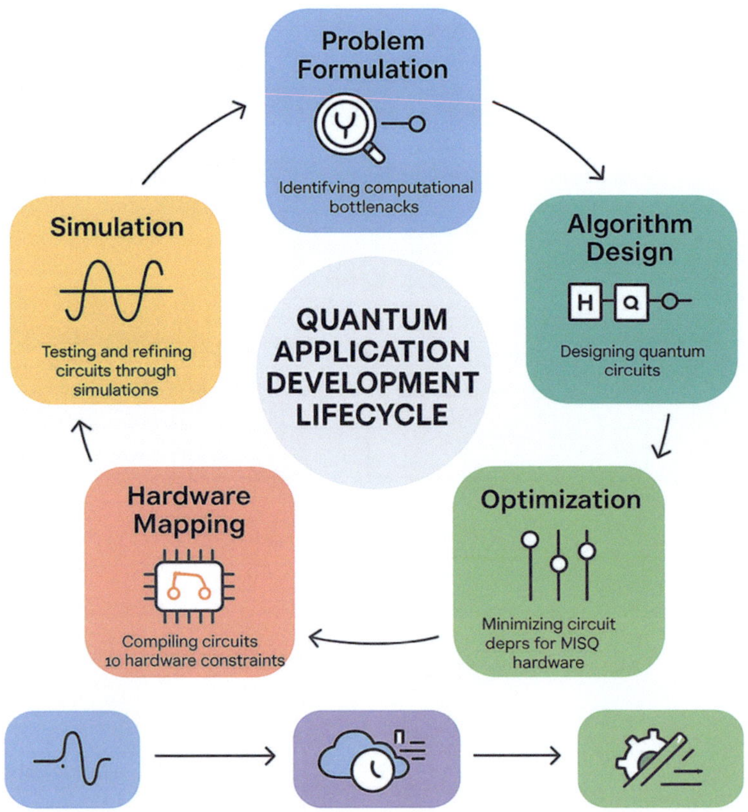

Fig. 15.4 Quantum computing applications development lifecycle [12]

- *Simulation*—Test and refine quantum circuits through software simulations to verify expected outputs. Tune circuit parameters before hardware execution.
- *Optimization*—Minimize the circuit depth and number of gates required using optimization frameworks to make it executable on NISQ hardware.
- *Hardware Mapping*—Compile and map quantum circuits to available hardware connectivity and gate constraints. Insert swap mappings to move qubits.
- *Job Scheduling*—Schedule execution jobs that account for qubit availability, priority, and queue depth on quantum cloud platforms.
- *Result Analysis*—Collect job outputs and log metrics. Analyze the results to validate algorithm performance on real hardware.
- *Error Mitigation*—Identify noise channels causing errors and use techniques like zero noise extrapolation to improve results—iterate algorithm refinements.

Mastering this lifecycle is key to unlocking business value from quantum techniques in real-world contexts. Abstractions and automation will enhance developer productivity like classical software [12]. However, physics constraints on current quantum hardware necessitate special consideration for compilation, optimization, and error management during implementation.

15.3 Debugging and Testing Quantum Programs

The probabilistic nature of quantum programs makes debugging and testing inherently challenging [13]. Circuit outputs are non-deterministic across runs due to qubit state fragility and measurement randomness as described in Fig. 15.5. Bugs can also be highly hardware specific. Critical quantum programming testing approaches include:

- *Unit Testing*—Using simulators, isolate and test individual quantum circuit components and gates. Leverage frameworks that support parameterized, reusable test circuits.
- *Regression Testing*—Maintain testing suites to run on code changes. Compare outputs across code revisions during refactoring to detect introduced errors.
- *Hardware Emulation*—Debug implementation issues early by emulating target quantum hardware constraints like limited connectivity when running on simulators.
- *Assertion-Based Testing*—Embed test oracles and assertions within circuits to validate intermediate outputs and detect faults.
- *Quantum State Tomography*—Partially reconstruct quantum states mid-circuit by measuring from different bases to infer complete qubit state evolution.
- *Hardware Noise Characterization*—Characterize the noise and errors of target quantum processors to realistically simulate them during development. Model hardware imperfections.

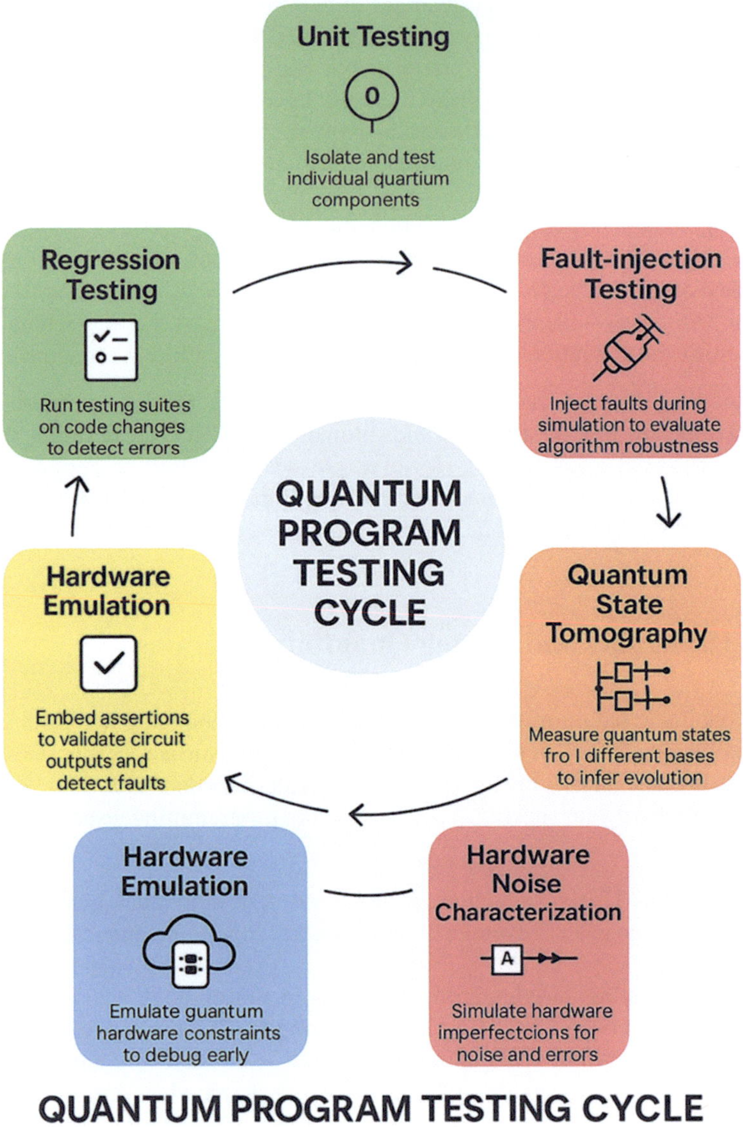

Fig. 15.5 Quantum computing program testing cycle

- *Fault-injection Testing*—Intentionally inject common hardware faults like dephasing or bit-flips during simulation to evaluate algorithm robustness.

Overall, developers should extensively leverage simulators paired with robust testing workflows to catch issues early before hardware execution [13]. This saves expensive quantum cycles and makes iterative debugging productive. Mature debugging tools and metrics tailored to quantum software will be vital.

Summary

This chapter provided a comprehensive introduction to programming quantum computers for business applications, covering languages, the development lifecycle, testing, and debugging. Mastering these practical skills in coding and optimizing quantum circuits, simulations, and cloud integrations will be essential to harnessing the advantages promised by quantum computing. For enterprises, cultivating in-house quantum software development expertise across their IT teams will catalyze innovation. Overall, quantum programming opens exciting opportunities and challenges for organizations to lead in the emerging quantum era.

References

1. Cross, A. W., Bishop, L. S., Smolin, J. A., & Gambetta, J. M. (2022). open quantum assembly language. *ACM Transactions on Quantum Computing, 3*(2).
2. Di Matteo, O. (2024, Feb 14). *On the need for effective tools for debugging quantum programs.* arXiv:2402.09547.
3. Efthymiou, S., Ramos-Calderer, S., Bravo-Prieto, C., Pérez-Salinas, A., & García-Martín, D. (2022). Qibo: A framework for quantum simulation with hardware acceleration. *Quantum Science and Technology.*
4. Häner, T., Steiger, D. S., Svore, K., & Troyer, M. (2018). A software methodology for compiling quantum programs. *Quantum Science and Technology.*
5. Liu, Y., Pei, H., Huang, L., Yin, B., & Cai, K.-Y. (2024). Automatic repair of quantum programs via unitary operations. *ACM Transactions on Software Engineering and Methodology, 33*(6), 154.
6. Qutes Team. (2025). *Qutes: A high-level quantum programming language....* arXiv.
7. Ramalho, N. C. L., de Souza, H. A., & Chaim, M. L. (2024). Quantum software engineering: Challenges and methodologies. *SCIRP.*
8. Rovara, D., Burgholzer, L., & Wille, R. (2024, Dec 16). *A framework for debugging quantum programs.* arXiv:2412.12269.
9. Rovara, D., Burgholzer, L., & Wille, R. (2024, Dec 18). *Automatically refining assertions for efficient debugging of quantum programs.* arXiv:2412.14252.
10. Sheladiya, Z. (2023). *Top 5 quantum programming languages to learn in 2023.* LinkedIn.
11. Springer Survey. (2025). *Quantum software engineering and quantum software development.*
12. Wiley Special Issue. (2024). *Quantum software development lifecycle.*
13. Fingerhuth, M., Babej, T., & Wittek, P. (2018). Open source software in quantum computing. *PLOS ONE.*

Chapter 16
Realizing Quantum AI Projects

The previous chapters introduced foundational concepts across quantum computing, machine learning, and software development. However, bridging theory to practice remains challenging. This concluding chapter guides planning and executing quantum AI projects in enterprise contexts.

Practical considerations like assessing quantum readiness, selecting feasible use cases, managing project risks, and building internal skills are discussed. A staged roadmap is presented to integrate quantum techniques into business workflows in a phased manner. This chapter equips executives, technology leaders, and project managers with best practices and advice for catalyzing value from quantum AI investments which has been represented in Fig. 16.1.

16.1 Steps for Integrating Quantum AI in Business Processes

Quantum computing opens possibilities to reimagine business processes by integrating quantum techniques like optimization, simulation, and AI into workflows for greater efficiency, insights, and automation [1]. However, open-ended application exploration is impractical as shown in Fig. 16.2.

A structured approach is recommended:

1. **Opportunity Assessment**

    - Map core business processes and pain points that could benefit from quantum techniques based on complexity, scale, and analytical needs.
    - Shortlist processes where classical methods are hitting limitations today, with biggest potential for quantum advantage.

M. Singh and B. S. Rawal, *Quantum Minds: Exploring Quantum Computing and AI Synergy*, https://doi.org/10.1007/978-981-95-2498-3_16

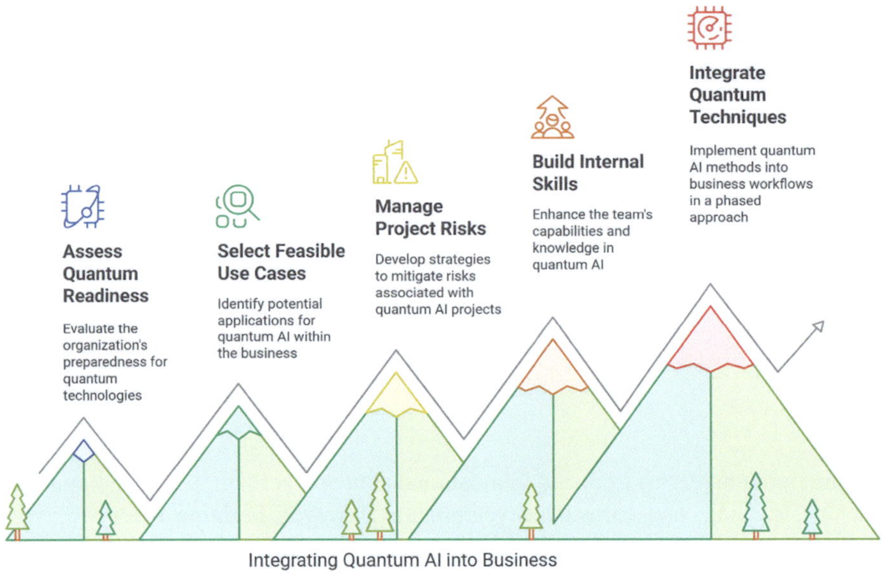

Fig. 16.1 Quantum AI integration in business

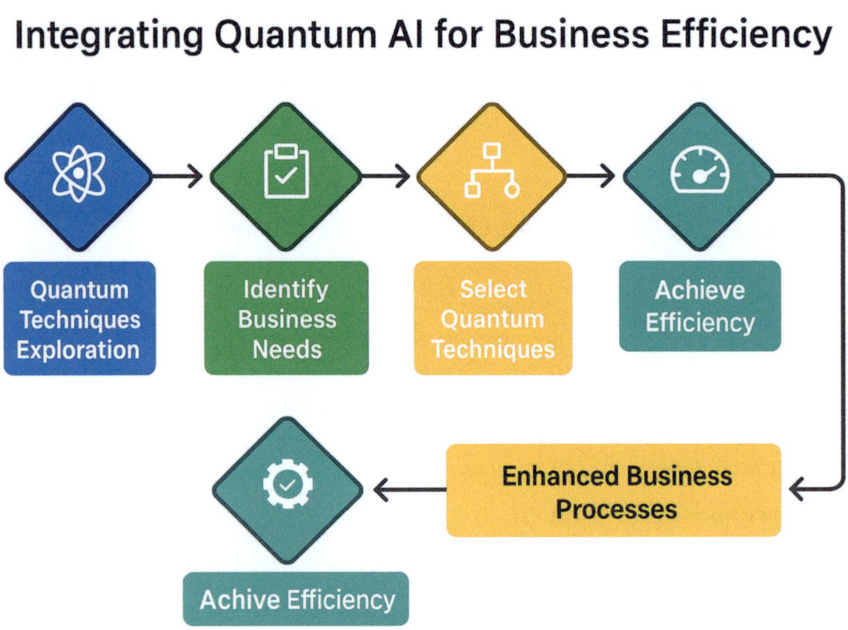

Fig. 16.2 Quantum AI for business efficiency

2. **Use Case Definition**
    - Detail specific subtasks and steps within priority processes that quantum approaches can target.
    - Frame use cases with concise problem definitions, success criteria, and required technical capabilities.
3. **Solution Scoping**
    - Research quantum algorithms, software libraries, and cloud services applicable to target use cases.
    - Estimate the effort, skills, and costs required for initial prototypes.
    - Secure executive sponsorship and budget.
4. **Proof of Value**
    - Develop a minimally viable solution for the use case on quantum simulators and cloud platforms.
    - Demonstrate improved performance over classical baselines on representative data samples.
5. **Pilot Deployment**
    - Engineer end-to-end hybrid classical-quantum workflow with interfaces to business systems.
    - Run controlled tests on production data and processes to validate advantages.
6. **Transition to Production**
    - Based on pilot results, refine the solution and expand integration for enterprise-wide production deployment.
    - Build supporting processes like monitoring, backups, and continuous improvement.

This staged approach allows quantum computing power to be focused on real business needs while managing project risks and skill requirements. Returns can be showcased incrementally to justify further investment [2].

16.2 Assessing Quantum AI Feasibility

Not all business use cases are suited for early stage quantum implementations. Rigorously assessing feasibility upfront is vital before committing significant investments [3]. It's represented in Fig. 16.3. The key criteria include:

- ***Problem ComplexityComplexity***—Quantum techniques provide an advantage for problems intractable to classical supercomputers. Simple analytics tasks offer limited returns.

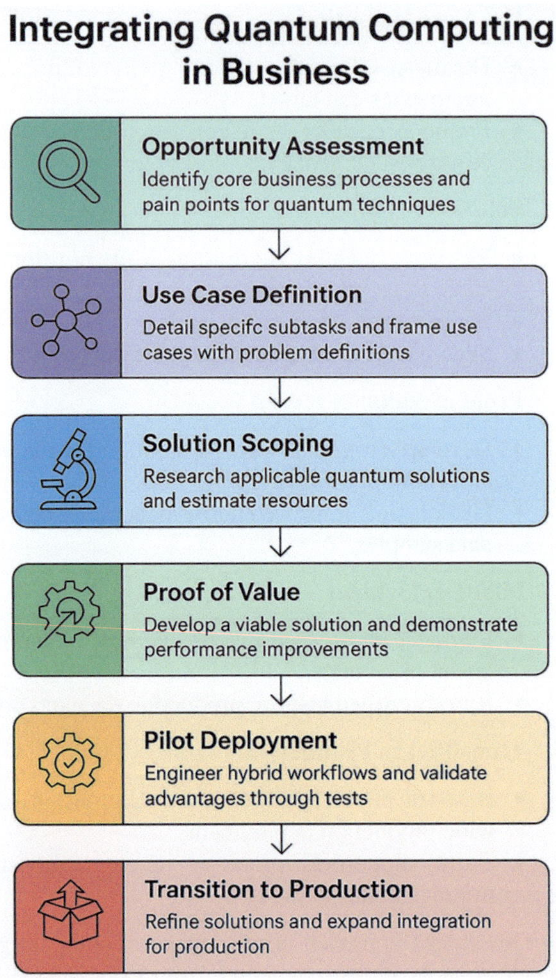

Fig. 16.3 Quantum computing integration in business

- ***Data Volume***—Quantum processors today have limited qubit counts. Use cases without exponential data growth can fit classical hardware.
- ***Hybridization Requirements***—Solutions requiring tight integration between quantum and classical systems are more complicated in the near term.
- ***Performance Advantage***—The advantage over classical techniques must be significant to warrant the overhead of quantum programming and infrastructure.
- ***Business Value Impact***—Focus on use cases that provide a disproportionate competitive advantage if successfully implemented.
- ***Solution Maturity***—The quantum algorithm must be sufficiently developed and tested to be viable for business needs.

Evaluating these factors will help identify quantum AI projects with maximum return potential within organizational constraints. Teaming business and technical experts is vital to perform credible feasibility analysis [4].

16.3 Overcoming Challenges in Quantum Adoption

Despite the promising potential, adopting quantum computing in business faces obstacles that must be navigated:

- **Immature Technology**—Production quantum systems are still emerging. Algorithms require refinement for business applications. Patience is needed to ride the maturation curve.
- **Specialized Expertise**—Deep quantum skills remain rare, and building capabilities are complex. Partnering with external experts can help address gaps.
- **Cultural Inertia**—Entrenched classical technology investments and workflows pose cultural resistance. Senior leadership must actively champion change.
- **Uncertain ROI**—Because it is a novel technology, reliable return on investment estimates is difficult to obtain. Pilots and incremental adoption are advisable.
- **Flexible Mindsets**—Quantum computing compels rethinking problems fundamentally differently. People must adapt to unfamiliar quantum modes of thought.
- **Vendor Fragmentation**—The quantum ecosystem has many disjointed toolsets. Integrating end-to-end solutions is challenging currently.
- **Emerging Best Practices**—Industry playbooks in applications, programming, and infrastructure are still maturing. Shared learning should be leveraged.

While challenging, none of these barriers are insurmountable. Other disruptive technologies like AI have also faced similar teething issues before reaching inflection points [5]. Developing expertise and experimentation today will best position organizations to capture quantum opportunities as depicted in Figs. 16.4 and 16.5.

16.4 Quantum AI Implementation Roadmap

The volume of hype surrounding quantum computing necessitates a measured, milestone-driven approach as shown in Fig. 16.6.

The following staged roadmap allows enterprises to build quantum capabilities responsibly:

- **Stage 1—Educate and Explore** [6]:
  - Promote quantum literacy across the organization.
  - Evaluate business use cases for quantum relevance.
  - Experiment with public cloud services for experience.

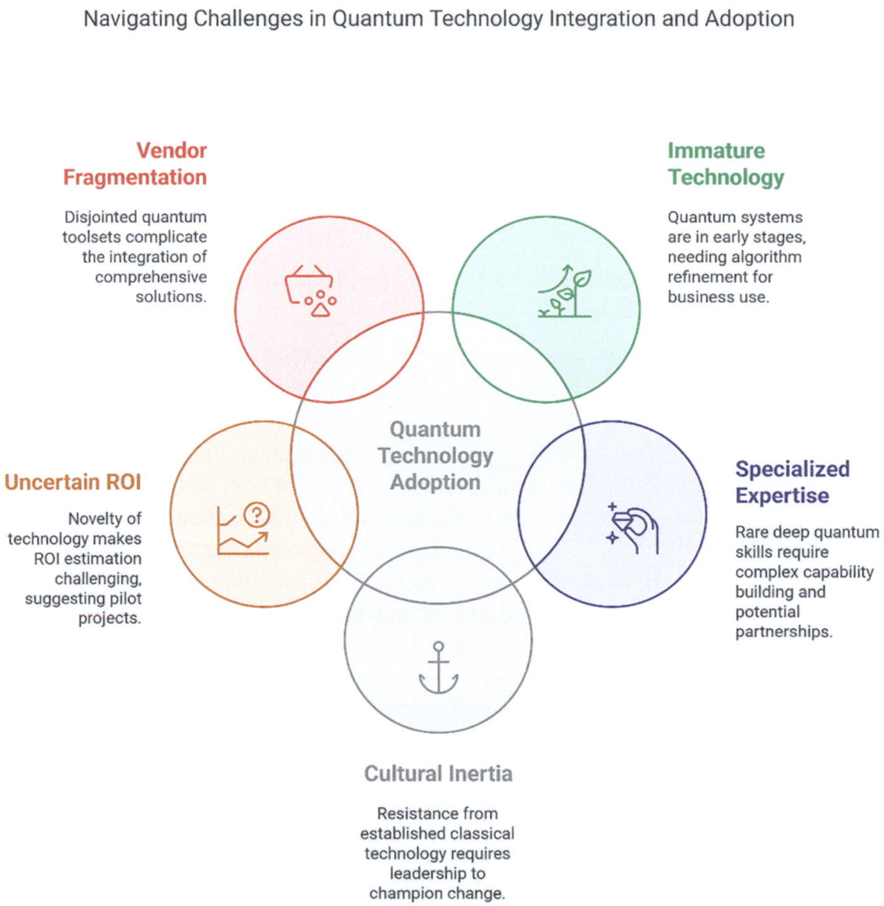

Fig. 16.4 Challenges in integration of quantum computing and it's adoption

- **Stage 2—Architect and Prototype** [7]:
  - Train specialized technical talent.
  - Design solution architecture for priority use case(s).
  - Build proof of concept on quantum simulators.
- **Stage 3—Demonstrate Value** [8]:
  - Realize minimal viable solutions on quantum hardware via the cloud.
  - Show tangible improvements over classical baselines.
  - Make case for operational pilots by showcasing value.
- **Stage 4—Scale and Integrate** [9]:
  - Operationalize end-to-end quantum workflows.
  - Identify ancillary processes like monitoring and backups.

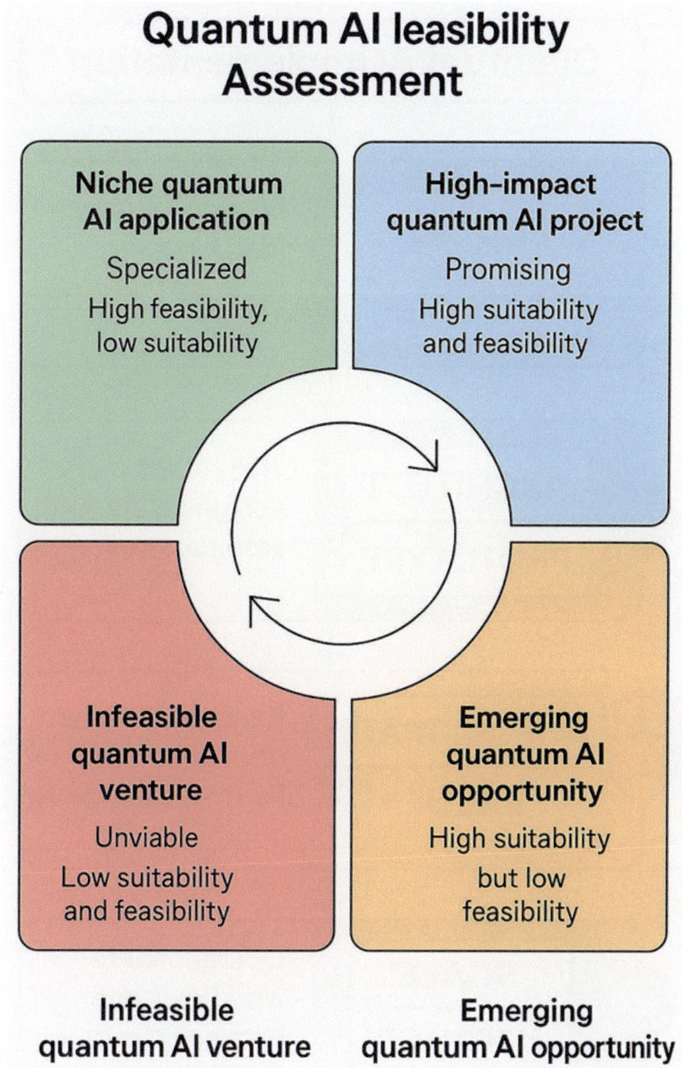

Fig. 16.5 Quantum AI feasibility assessment

– Expand integration and usage across the enterprise.

This roadmap allows an organization to incrementally advance their quantum AI capabilities while managing risks. Leadership commitment to sustained capacity building is vital to reap rewards.

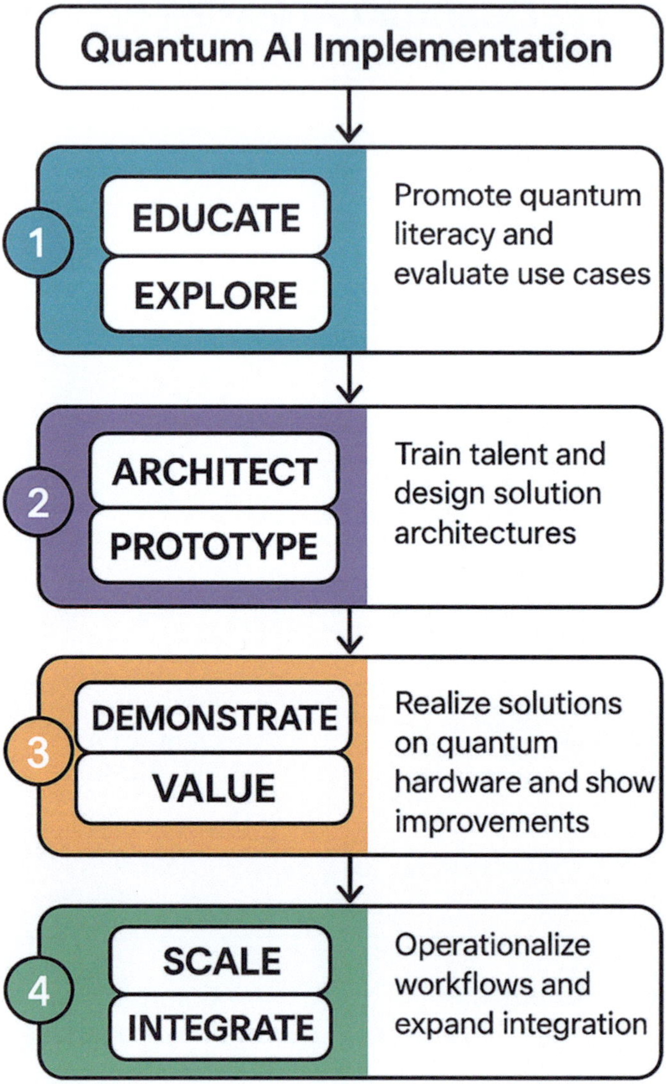

Fig. 16.6 Quantum AI roadmap

Summary

This chapter presented a comprehensive guide to planning and executing quantum AI projects tailored for business environments. Quantum computing is transitioning from pure science fiction into a domain with tangible commercial applications. However, hype must be counterbalanced with pragmatism. Following structured approaches outlined in this chapter will enable enterprises to extract real advantages from quantum techniques in the emerging hybrid quantum–classical computing paradigm.

References

1. Mueck, M. (2017). Quantum software engineering: a smart approach to quantum computing. *Software: Practice and Experience, 47*(11), 1575–1582.
2. Humble, A., & McMahon, D. (2020). How to start out in quantum computing as an enterprise. *IEEE Computer, 53*(9), 61–71.
3. Fu, X., Riese, A., Zhao, J., Majumder, P., Sau, R., & Anis, M. Z. (2022). Quantum programming environments, libraries and software stacks. *ACM Computing Surveys (CSUR), 55*(4), 1–34.
4. Dunjko, V., & Briegel, H. J. (2018). Machine learning & artificial intelligence in the quantum domain: A review of recent progress. *Reports on Progress in Physics, 81*(7), 074001.
5. Zschech, P., Heinz, C., Janiesch, C., & Vollmer, A. (2024). Modern finance through quantum computing—A systematic literature review and research agenda. *Financial Innovation, 10*, 66. https://doi.org/10.1186/s40854-024-00668-7.
6. Acampora, G., Ambainis, A., Ares, N., Banchi, L., Bhardwaj, P., Kockum, A. F., et al. (2025). *Quantum computing and artificial intelligence: status and perspectives.* arXiv:2505.23860.
7. Artkai. (2024). *A roadmap for integrating AI into enterprise software systems.* Artkai Blog.
8. McKinsey & Company. (2021). *Quantum computing use cases are getting real—What you need to know.* McKinsey Digital.
9. Oliver Wyman & SFF. (2025). *AI and quantum: Operationalizing advanced technologies in financial services.* Oliver Wyman Report.

Part V
Looking Ahead

The previous parts provided a comprehensive introduction to quantum computing, quantum machine learning, and their applications for business professionals. However, we are still in the very early stages of the quantum AI revolution. Tremendous progress is anticipated in the years and decades ahead. This concluding part gazes forward to envisioning the landscape as quantum technologies mature from narrow niches into a generalized computing paradigm.

Business leaders should actively track scientific and commercial advances in the quantum ecosystem. While timelines are uncertain, quantum capabilities will undoubtedly transform industries. Strategic planning must account for the disruptive potential of quantum techniques maturing into their full potential. This final part aims to equip readers with a forward-looking perspective on the future quantum AI landscape.

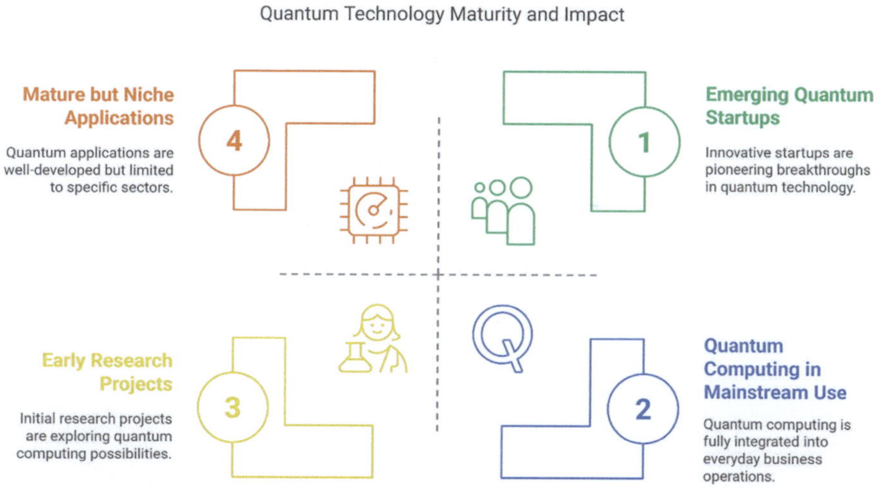

Fig. 5 Quantum computing looking ahead

Chapter 17
Getting Involved in Quantum Computing

Quantum computing is an exciting emerging field with many opportunities for involvement. This chapter explores ways to engage with quantum computing communities, contribute to open source, learn quantum programming, and pursue higher quantum education and career opportunities till March 2025.

17.1 Joining Online Quantum Communities

Many online forums and groups provide ways to connect with the quantum computing community to discuss developments, learn, and find job opportunities [1–4]:

- **/r/quantumcomputing on Reddit**—Active subreddit covering latest news and breakthroughs.
- **Quantum Computing StackExchange**—Q&A site for quantum computing questions.
- **Quantum Computing Report Slack**—Invite-only Slack community of researchers and professionals.
- **Qiskit Slack Community**—Open Slack community managed by IBM Research.
- **DECIGO on Discord**—Open Discord for quantum programming discussions.
- **DiversityQ on Slack**—Focused on improving quantum diversity.
- **XRDS on Telegram**—Connecting early career researchers.

Conferences like Q2B also foster valuable niche communities. Engaging online helps build knowledge, relationships, and careers in quantum science.

M. Singh and B. S. Rawal, *Quantum Minds: Exploring Quantum Computing and AI Synergy*, https://doi.org/10.1007/978-981-95-2498-3_17

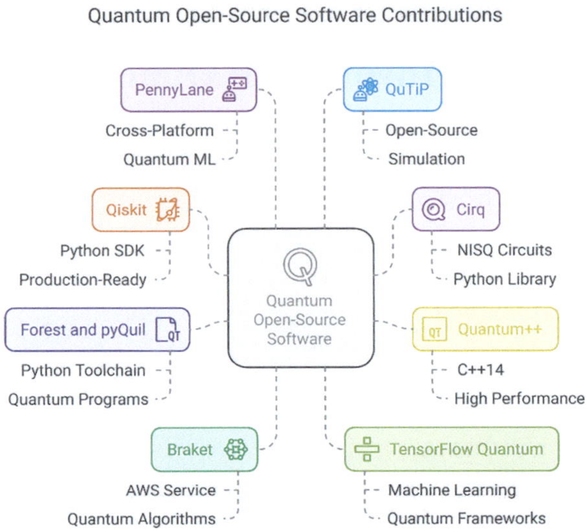

Fig. 17.1 Quantum open-source software contributions

17.2 Contributing to Quantum Open-Source Software

Much of quantum computing development occurs in open-source software projects. Contributions help advance the ecosystem's tooling and capabilities as shown in Fig. 17.1 [5–8]:

- Qiskit from IBM—Leading production-ready quantum SDK with Python.
- Cirq from Google—Python library for NISQ circuits.
- Quantum++—High-performance C++ 14 quantum computing library.
- Forest and pyQuil from Rigetti—Python toolchain for quantum programs.
- Braket from Amazon—Managed AWS service for running quantum algorithms.
- TensorFlow Quantum—Quantum machine learning frameworks.
- PennyLane—Cross-platform Python for quantum ML.
- QuTiP—Open-source Python for simulating quantum systems.

Contributions like documentation, tutorials, bug fixes, optimizations, and new features help improve these critical projects. GitHub is the predominant platform has represented in Fig. 17.2.

17.3 Learning Quantum Programming

Gaining quantum coding skills prepares for working in quantum computing. Many resources exist for learning [9–11] as per Fig. 17.3:

- IBM Quantum Challenge—Guided quantum coding challenges using Qiskit.

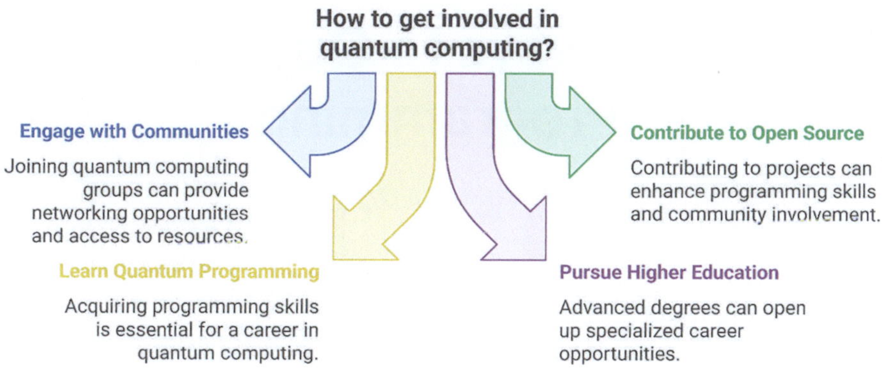

Fig. 17.2 Quantum computing involvement

- Qiskit Textbook—Interactive learning quantum software skills.
- Microsoft Quantum Katas—Hands-on quantum katas with the Q# language.
- QC Ware Meta Game—Gamified guided quantum coding tutorials.
- Brilliant Quantum Computing—Problem-based learning focused on theory.
- Coursera Quantum Courses—Structured online classes from universities.
- Coding quantum circuits, algorithms, and applications is the best way to cement concepts from theory. Languages like Python, Q#, and C++ feature quantum libraries.

17.4 Higher Education and Career Opportunities

- Specialized quantum education prepares for quantum careers [1]:
- Undergraduate—Physics and computer science concentrations in quantum information science provide foundations.
- Graduate Research—Pursuing master's and PhD studies in quantum computing grows expertise for research roles.
- Interdisciplinary—Cross-department quantum centers allow focusing across physics, CS, engineering, chemistry, and mathematics.
- Certificates—Professional graduate certificate programs offer quantum education for working professionals.

The quantum job market is rapidly expanding, and qualified talent is scarce. Opportunities exist across hardware development, software engineering, algorithm research, education, management, and more [12–15]. Entry-level roles benefit from basics in linear algebra, probability, CS algorithms, and physics. Specializing early provides an advantage. The quantum computing field offers many exciting career directions (Fig. 17.4).

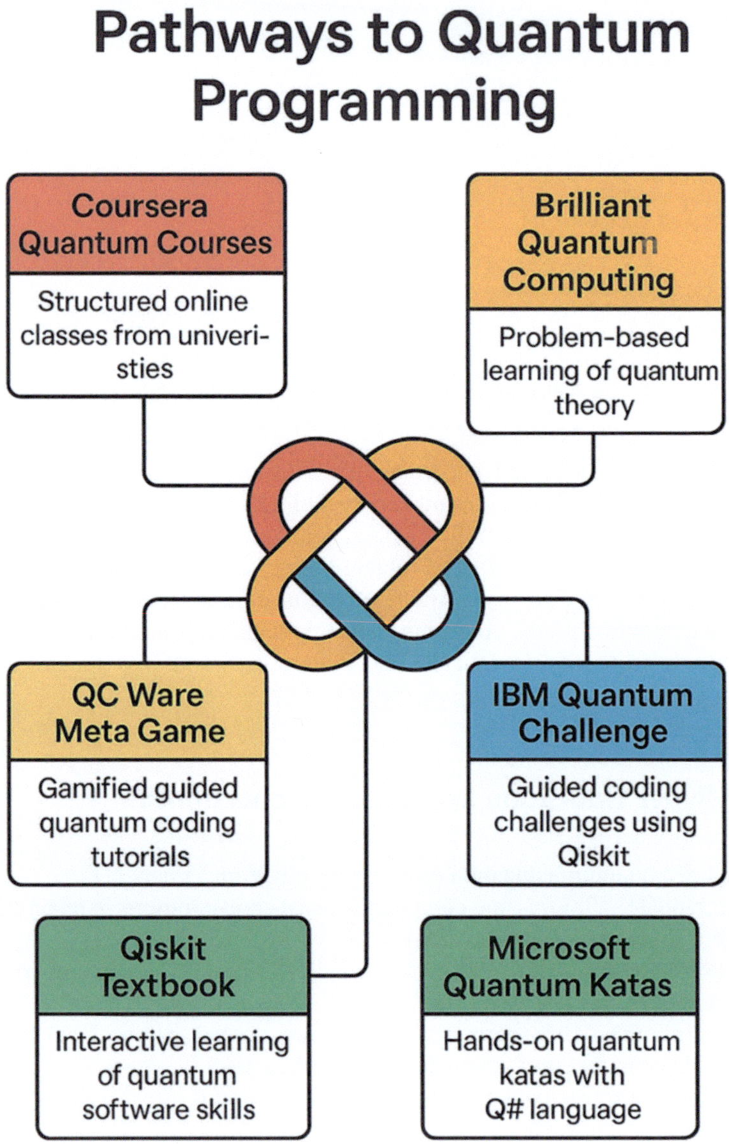

Fig. 17.3 Quantum programming pathways

Fig. 17.4 Quantum computing opportunities

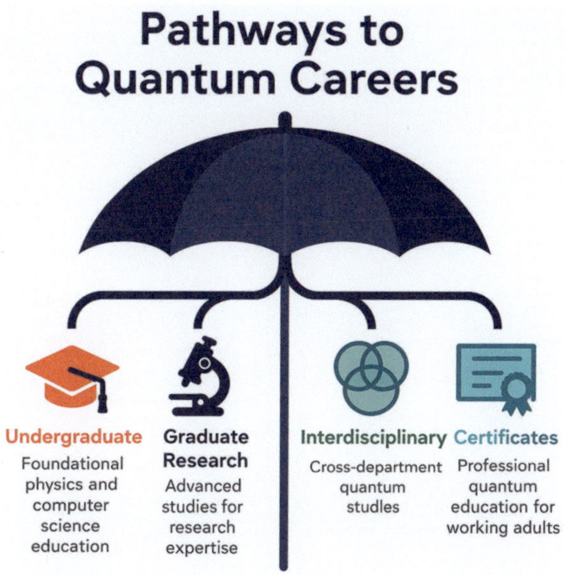

References

1. Gajjar, D. (2025, Feb 20). *How open source is making quantum computing accessible to everyone*. OpenSourceForU.
2. Quantumzeitgeist. (2024). *Ideas to become a quantum programmer: Join the quantum subreddit, participate in hackathons.*
3. IBM Quantum. (2025). *IBM Quantum Community: network of 600K+ members, Qiskit advocates.*
4. Reddit user. (2021). *Online community for quantum grad students?* Reddit thread on spiritual support and collaboration.
5. QOSF. (2025). Quantum computing mentorship program: contribute to Qiskit, Cirq, etc.
6. StackExchange user. (2024). How can a beginner start contributing to Qiskit/Cirq?
7. OpenSourceForU. (2025, Feb 20). *Open source initiatives democratizing quantum computing.*
8. Shammah, N., et al. (2023). *Open hardware solutions in quantum technology*. arXiv.
9. Mendoza-Cortés, J. L., et al. (2025). *Machine learning and quantum computing guide for humans*. Open-access e-book.
10. PennyLane team. (2024, Mar). *From classical to quantum software development*. PennyLane Blog.
11. Wootton, J. R., Harkins, F., Bronn, N. T., et al. (2020). *Teaching quantum computing with an interactive textbook ("Learn Quantum Computation using Qiskit")*. arXiv.
12. Kaur, M., & Venegas-Gomez, A. (2022). *Defining the quantum workforce landscape: review of global quantum education initiatives*. arXiv.
13. Business Insider. (2025, Apr 26). *Quantum computing gears up for its "ChatGPT Moment" & a potential talent shortage.*
14. NerdWerk. (2025). *Global career paths in quantum computing companies.*
15. UChicago Professional. (2024). *High-paying quantum tech jobs growing fast—Grad degree not required.*

Chapter 18
The Future of Quantum Computing

Realizing quantum computing's full potential to impact society requires surmounting significant challenges related to scaling quantum systems, achieving robust error correction, building quantum networks, and commercialization. This chapter analyzes the outlook and technical landscape for advancing quantum computing on these pivotal fronts.

18.1 Scaling Up Quantum Systems

Building large-scale, reliable quantum computers with high qubit counts remains an immense engineering challenge [1] as represented in Fig. 18.1.

18.2 Key Areas for Scaling Focus on

- **Increasing Qubit Number**
  All mainstream quantum computing technologies must overcome hurdles to scaling qubit count while maintaining or improving fidelity and lifetimes. For superconducting qubits, this involves advances in fab processes, materials, and circuit design. Trapped ions face challenges expanding trap arrays. Quantum photonics seeks lossless components [2]. Qubit count benchmarks quantum volume as shown in Fig. 18.2.
- **Improving Qubit Connectivity**
  Expanding a qubit of connectivity facilitates expressing more complex algorithms and entanglement. Superconducting chips require increasingly dense circuit wiring for control systems. Trapped ions can conditionally transport qubits.

M. Singh and B. S. Rawal, *Quantum Minds: Exploring Quantum Computing and AI Synergy*, https://doi.org/10.1007/978-981-95-2498-3_18

Fig. 18.1 Scaling up quantum systems

Fig. 18.2 Increasing qubit number

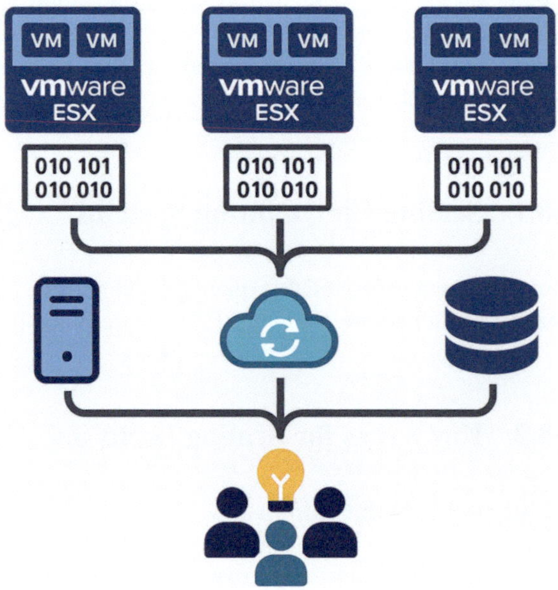

Photonic systems aim to lower photon loss between components [3]. Dense connectivity unlocks algorithmic scaling.

- **Enhancing Gate Fidelities**
  High-fidelity quantum logic gates are essential for multi-qubit circuit depth and complexity. This requires fine-tuned qubits with long coherence times and high-precision control electronics/lasers [4]. Gate error rates below a threshold enable fault tolerance.
- **Raising Clock Speed**

Faster qubit control cycles multiply algorithmic operations within coherence lifetimes. Rapid spin resonance manipulations, fast laser pulses, and low-latency classical electronics/FPGAs accelerate sequencing. Fast clock rates shrink algorithm runtimes [5].

These four factors contribute to the conditions for achieving supreme quantum computational power.

18.3 Quantum Error Correction

The noise and fragility of qubits lead to decoherence and gate errors that will make large-scale quantum computation impossible without error correction [6]. Quantum error-correcting codes embed logical qubits redundantly across many physical qubits to detect and correct errors by majority vote.

For example, the surface code uses a 2D lattice of qubits with code stabilizers that continually detect phase and bit-flip errors. Any single physical qubit error can be corrected by the logical state encoded in neighboring qubits.

Repeating error detection and running error-prone operations fault-tolerantly, arbitrarily reliable logical qubits can be maintained given a physical error rate below a threshold value. This exact threshold value is still being optimized, but leading estimates range from 0.1 to 1% [7]. Various quantum error-correcting encodings are known, including:

- Surface codes using lattice grids of physical qubits.
- Concatenated codes recursively encode logical qubits.
- Topological codes creating nonlocal protection.
- Subsystem codes with simplified syndrome extraction

The space overhead for functional error-corrected qubits remains high. Logical qubits may require millions of noisy physical qubits. Reducing this overhead is an active area of research.

Quantum error correction will be essential for realizing scalable, fault-tolerant quantum computation necessary to run meaningfully large algorithms like Shor's.

18.4 Quantum Networking and Communications

Connecting multiple quantum processors via quantum entanglement will enable distributed applications like clock synchronization, consensus, blinded computation, finance, secret sharing, and cryptography [8]. Networking qubits require (Fig. 18.3):

- Generating long-lived stationary entangled states between network nodes through parametric down-conversion or trapped ions/atoms.

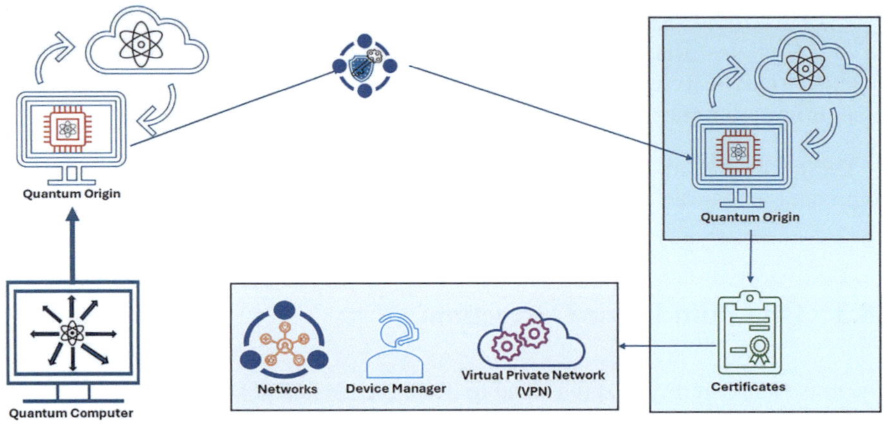

Fig. 18.3 Quantum networks

- Storing qubits at network nodes in quantum memory with long coherence times.
- Establishing inter-conversion between flying and stationary qubits.
- Faithfully transmitting flying qubits between locations through channels like free-space QKD or fiber optics.
- Repeating/routing qubits with quantum repeaters to extend range.
- Running quantum teleportation protocols and superdense coding circuits across the network.

With high-fidelity stationery and flying qubits entangled at large scales, truly distributed quantum computing, metrology, and communication have become possible [9].

18.5 Commercialization Outlook and Challenges

In recent years, significant investment and remarkable physics advances have occurred across quantum computing. But real-world commercial adoption requires crossing some remaining hurdles [10]:

- Achieving sufficient qubit number, fidelity, and algorithmic quantum advantage.
- Improving programmability, workflows, and abstraction layers higher than assembly.
- Building integrations bridging the classical-quantum divide.
- Cultivating industry domain expertise and use-case discovery.
- Creating supporting ecosystem infrastructure and service layers.
- Addressing volatility and speculation risks in the quantum investment landscape.
- Achieving ROI clarity and quantum business model viability.
- Securing reliable access to affordable quantum hardware resources.

With thoughtful collaboration across research, industry, and government, these challenges are steadily being overcome. As quantum computing matures over the next decade and beyond, its commercial promise should be fully unleashed—introducing disruptive new computational capabilities.

References

1. Zwolak, C., et al. (2025). Automated calibration of scalable quantum-dot qubit arrays. *Physical Review Letters.*
2. SpinQ. (2025, Feb). *Scalable quantum systems: The path to quantum advantage.* SpinQ Blog.
3. IBM Quantum. (2024). *Development roadmap: Building a 100,000+ qubit system.* IBM Quantum Roadmap.
4. National Academies of Sciences, Engineering, and Medicine. (2019). *Quantum computing: progress and prospects.* The National Academies Press.
5. Cai, Z., et al. (2022*). Quantum error mitigation.* arXiv:2210.00921.
6. Google Quantum AI. (2023). Surface code threshold surpassed: Real-time quantum error correction. *Nature.*
7. IBM Quantum. (2025, June 10). *IBM lays out clear path to fault-tolerant quantum computing by 2029.* IBM Quantum Blog.
8. Wehner, S., Elkouss, D., & Hanson, R. (2018). Quantum internet: A vision for the road ahead. *Science, 362*(6412), eaam9288.
9. McKinsey & Company. (2021). *Quantum computing: Use cases are getting real—What you need to know.* McKinsey Digital.
10. Barron's. (2025, June). *IonQ positioned to become a leader in quantum computing and networking.*

Chapter 19
Quantum AI Trends and Future Outlook

Quantum computing has progressed rapidly from a pure science fiction dream decades ago into early but influential prototype systems today. However, we are still far from unlocking its full disruptive potential across industries [1]. The field remains active, with discoveries continually expanding the horizons of the possible as shown in Fig. 19.1.

This section highlights anticipated technology milestones shaping quantum's continued march to business relevance over the next decade and beyond.

19.1 Quantum AI in the Era of Quantum Advantage

Today's era of Noisy Intermediate-Scale Quantum (NISQ) computing offers glimpses into the quantum future has represented in Fig. 19.2. But fully unleashing the power of quantum will require crossing key tipping points [2–6]:

- *Quantum Volume ~ 4096*—At this scale with around 50–60 qubits, algorithms like quantum chemistry simulations become viable but still limited. Cloud platforms are approaching this milestone.
- *Quantum Volume ~ 1 million*—Quantum circuits finally exceed classical supercomputer capabilities. This is the threshold for claimed quantum advantage over classical methods.
- *Fault Tolerant Qubits*—Quantum error correction allows practically large circuit depths. This unlocks the full potential of quantum algorithms. Still 10+ years away.
- *Quantum Neural Networks*—With fault tolerance, training complex quantum neural networks that outperform classical AI becomes practical. This is a game-changer for AI.

M. Singh and B. S. Rawal, *Quantum Minds: Exploring Quantum Computing and AI Synergy*, https://doi.org/10.1007/978-981-95-2498-3_19

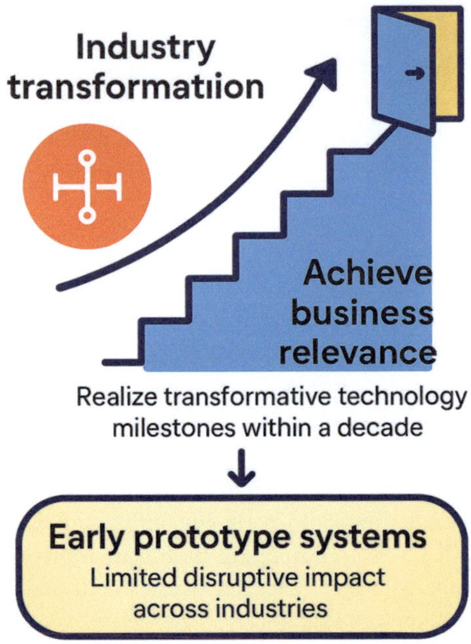

Fig. 19.1 Unlock quantum computing's opportunities

- *Quantum Internet*—Connecting quantum processors over entangled communication networks promises new distributed computing paradigms and allows quantum cloud services.

These milestones will not arrive as sharp transitions but rather as critical capabilities gradually maturing over the next decade.

Business leaders should track technical progress to identify emerging opportunities at each stage while maintaining realistic expectations. Quantum advantage will first materialize for specific applications rather than all workloads. Due to hardware constraints, hybrid quantum–classical capabilities will dominate in the near term. Nonetheless, the direction is clear, we are accelerating into a quantum future [7].

19.2 Anticipated Transformations in Management Practices

As quantum technologies mature in capability and commercial availability, their business impact will greatly expand beyond niche proof-of-concept use cases. The paradigm shift will occur once quantum techniques permeate core management disciplines and processes [8, 9]. Potential transformations include:

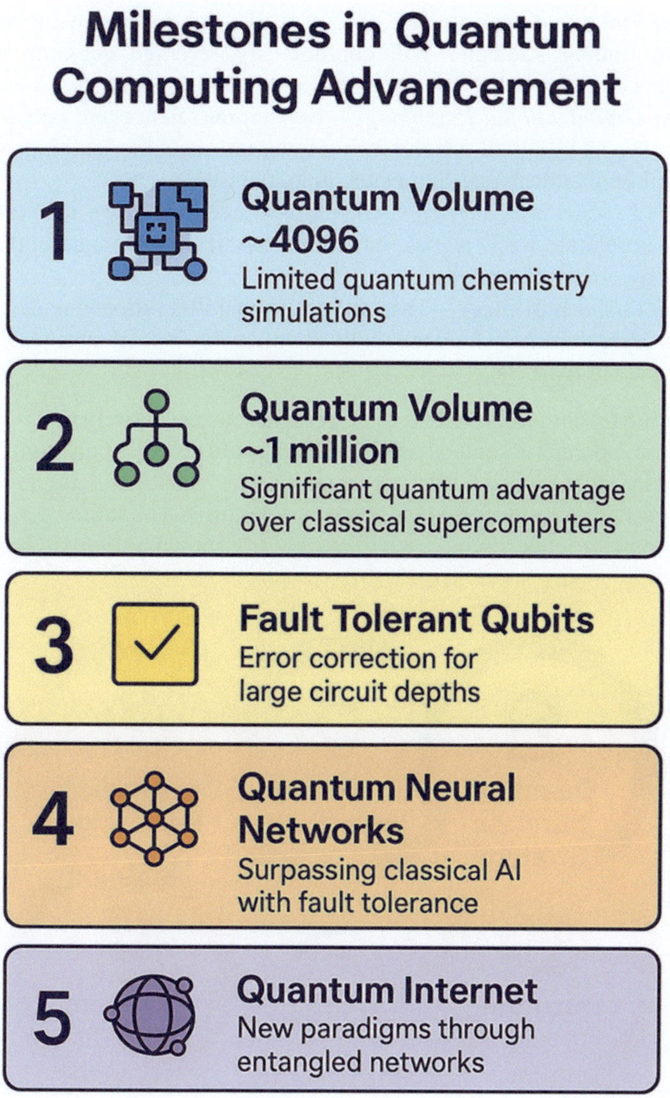

Fig. 19.2 Quantum computing advancements

- *Planning with Quantum Foresight*—Strategic planning and forecasting will tap into powerful quantum predictive analytics trained on massive datasets and simulations. This allows for more insights and scenario modeling for long-term decision-making.
- *Quantum-Powered Logistics*—Supply chains, shipping fleets, and warehouses will be dynamically optimized using compact quantum optimizers and simulators. This enables responsive edge logistics.

- *Quantum FinOps*—Quant finance processes like risk management, fraud detection, and trading strategies will continuously leverage quantum computing advantages on live market data for split-second decision-making.
- *Quantum Organizational Psychology*—Behavioral science and people analytics will uncover new human insights using quantum machine learning on multidimensional human decisions and psychology datasets.
- *Quantum Product and Service Innovation*—Accessing large quantum simulators will accelerate R&D across materials, chemicals, drugs, and other product domains by orders of magnitude.
- *Quantum Business Strategy*—The exponential parallel processing capabilities of quantum computers may fundamentally alter how businesses formulate strategies and compete in markets.

As quantum techniques permeate management science, entirely new paradigms of value creation, operations, and decision-making could emerge. Quantum computing may engender a metamorphosis across industries analogous to how classical computing reshaped the business landscape decades prior. The strategic opportunities are tremendous for leaders who can capitalize early shown in Fig. 19.3.

Fig. 19.3 Quantum computing transformation in business

19.3 Ethical Considerations in Quantum AI

The tremendous powers unlocked by quantum AI also compel deep ethical reflection on its responsible use for business and society [10]. Some key considerations include:

- *Avoiding algorithmic* biases and ensuring fairness should be proactive goals when deploying quantum machine learning systems.
- *New capabilities like quantum sensing or decryption* could infringe on reasonable privacy expectations without checks. Policy needs to align with emerging realities.
- *The environmental footprint of quantum data centers* will grow with adoption. Sustainability must be emphasized in quantum infrastructure.
- *Quantum cyber risks like codebreaking* must be weighed when offering new services. Post-quantum encryption will likely become mandatory.
- *Without diligence*, access to powerful quantum computers could overly concentrate economic advantage if too centralized in large corporations. Public access mechanisms could promote equity.
- *Quantum AI* may disrupt employment markets as workflows get redefined and automated. Proactive policies for workforce adaptation are prudent.

Overall, quantum technologies have ethical dilemmas and risks alongside advantages. Business leaders must steer these formidable capabilities toward positive societal ends. Corporate social responsibility should shape strategic quantum initiatives [11]. Wise governance and democratization will help quantum AI fulfill its constructive potential navigated in Fig. 19.4.

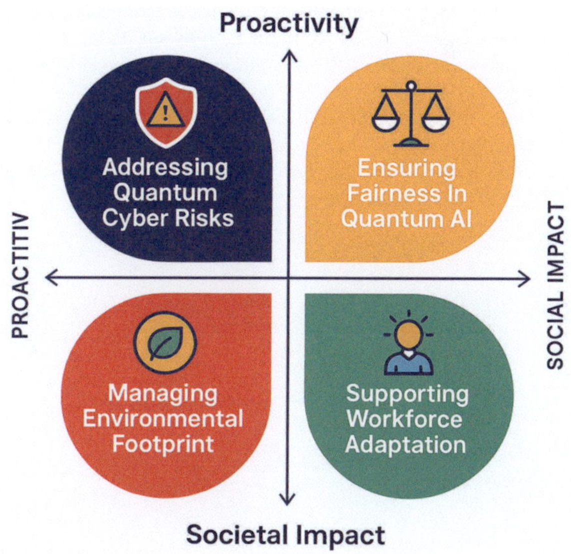

Fig. 19.4 Quantum computing challenges

Summary

This concluding part provided a forward-looking analysis of the quantum AI landscape as capabilities mature over the next decade. While predictions are complex, we sit at history's threshold today. Quantum computing heralds one of the most significant paradigm shifts since the emergence of classical electronic computers. Early movers who develop expertise and explore applications today will be firmly positioned to surf this seismic business transformation quantum technologies will catalyze across industries.

References

1. Svore, K. M., Geller, M., Troyer, M., Azariah, J., Granade, C., Heim, B., Wiebe, N., et al. (2022). Qubits for business: The opportunities and challenges quantum computing presents for companies today. *ACM Transactions on Quantum Computing, 3*(1), 1–34.
2. How, M.-L., & Cheah, S.-M. (2024). Forging the future: Strategic approaches to quantum AI integration for industry transformation. *AI*, *5*(1), 290–323. https://doi.org/10.3390/ai5010015
3. Lynch, D. (2025). Quantum AI trends. Quantum AI Trends»Empathy First Media.
4. Sfoglia, P. (2025). The perfect storm: Quantum computing AI and CMMC. VIEWPOINT: The Perfect Storm: Quantum Computing, AI and CMMC
5. Bano, M., Ali, S., & Zowghi, D. (2024, Oct). *Envisioning diversity and inclusion for quantum software engineering*. arXiv.
6. Acampora, G., et al. (2025). *Quantum computing and artificial intelligence: status and perspectives*. arXiv.
7. Castelvecchi, D. (2019). Quantum cloud computing. *Nature, 543*(7644), 159–161.
8. Mohseni, M., Read, P., Neven, H., Boixo, S., Denchev, V., Babbush, R., & Martinis, J. (2017). Commercialize quantum technologies in five years. *Nature News, 543*(7644), 171.
9. Olson, S. J., Cao, Y., Romero, J., Johnson, P. D., Dallaire-Demers, P. L., Sawaya, N. P., & Aspuru-Guzik, A. (2019). *Quantum information and computation for chemistry*. arXiv preprint arXiv:1706.05413.
10. Orús, R., Mugel, S., & Lizaso, E. (2019). Quantum computing for finance: Overview and prospects. *Reviews in Physics, 4*, 100028.
11. Preskill, J. (2018). Quantum computing in the NISQ era and beyond. *Quantum, 2*, 79.

Appendix A: Quantum Computing Concepts

- In-depth primer on superposition, entanglement, quantum measurements.
- Visual diagrams illustrate ambiguous quantum states.
- Technical introductions to various quantum logic gates (Hadamard, CNOT, SWAP, etc.) with example circuits.
  - Detailed explanations and circuit diagrams for key quantum algorithms:
  - Shor's algorithm for factoring large numbers.
  - Grover's algorithm for unstructured search.
  - Quantum phase estimation algorithm.
  - Quantum Fourier Transform for frequency analysis.
  - Quantum linear systems algorithm.
  - Descriptions of leading qubit modalities with schematics:
  - Superconducting qubits.
  - Trapped ion qubits.
  - Photonics qubits.
  - Topological qubits.
  - Explanations of quantum hardware components like cryogenics, controllers, and interconnects with architecture diagrams.
- Profiles of key quantum computing companies like IBM, Google, Microsoft, Rigetti, etc.

M. Singh and B. S. Rawal, *Quantum Minds: Exploring Quantum Computing and AI Synergy*, https://doi.org/10.1007/978-981-95-2498-3

Appendix B: Quantum Machine Learning

- Technical details on various quantum ML approaches with conceptual diagrams:
  - Quantum neural networks—perceptions, convolution, recurrent.
  - Quantum classifiers—quantum SVM, decision trees, k-nearest neighbors.
  - Quantum principal part analysis.
  - Quantum clustering algorithms.
  - Quantum regression models like quantum support vector regression.
- Mathematical formalism behind amplitude encoding, angle encoding, qubit encoding of classical data.
- Code examples showing key quantum ML techniques:
  - Data preparation and encoding.
  - Supervised learning workflows—quantum SVM, quantum neural networks.
  - Unsupervised learning—quantum PCA for dimensionality reduction.
  - Quantum ML using hybrid classical quantum workflows.
  - Practical hardware challenges in implementing quantum ML models on NISQ systems.
- Pathways to realizing quantum advantage for machine learning—quantum data, models, and hardware.

M. Singh and B. S. Rawal, *Quantum Minds: Exploring Quantum Computing and AI Synergy*, https://doi.org/10.1007/978-981-95-2498-3

Appendix C: Enterprise Use Cases

- Supply chain optimization using quantum annealing.
- Volkswagen—quantum algorithms for traffic flow optimization.
- Quantum machine learning for drug discovery—molecular interaction regression.
- Quantum risk simulation in financial services:
- Credit risk modeling with quantum Monte Carlo.
- Capital optimizations using quantum linear solvers.
- Portfolio stress testing using quantum generative models.
- Quantum natural language processing for enterprise search—pilot study.
- Quantum machine learning for predictive maintenance in manufacturing.
- Quantum anomaly detection for cybersecurity in telecom company.

M. Singh and B. S. Rawal, *Quantum Minds: Exploring Quantum Computing and AI Synergy*, https://doi.org/10.1007/978-981-95-2498-3

Appendix D: Quantum Software Development

- Detailed feature comparison of leading SDKs—Qiskit, Cirq, Amazon Braket, Microsoft QDK.
- Code examples demonstrating:
  - Quantum programming workflows—state preparation, operators, measurement.
  - Interfacing with quantum simulators and real hardware backends.
  - Quantum circuit optimization techniques—mapping, qubit routing, gate synthesis.
  - Error mitigation strategies—quantum error correction, noise modeling.
  - Guidelines for debugging, testing, simulating quantum programs.
  - Architecting scalable hybrid quantum-classical applications.
  - Quantum software development lifecycle—requirements, design, testing, deployment.
  - Addressing key challenges like hardware constraints, noise, decoherence.
  - Recommended software development practices for enterprise quantum projects.

M. Singh and B. S. Rawal, *Quantum Minds: Exploring Quantum Computing and AI Synergy*, https://doi.org/10.1007/978-981-95-2498-3

Glossary

- ***Quantum Computer***—An actual machine that uses qubits and quantum gates to perform computations. Requires isolating qubits and controlling interactions to maintain quantum coherence. Current examples include systems by IBM, Google, Rigetti.
- ***Quantum Computing Hardware***—The physical implementations and architectures of quantum computers. Some main approaches include superconducting qubits, trapped ions, and topological qubits. Each has its own advantages and challenges. Maintaining quantum coherence and interconnectivity as systems scale remains a key hardware challenge.
- ***Quantum Cloud Computing***—Services like Amazon Braket, IBM Quantum Experience, etc. allow accessing quantum computers remotely over the internet. Useful for testing algorithms and exploring quantum programming without requiring localized quantum hardware.
- ***Quantum Parallelism***—The ability of a quantum computer to evaluate a function for many different inputs simultaneously due to superposition. This massive parallelism enables quantum speedups.
- ***Quantum Volume***—A hardware-agnostic metric proposed by IBM to measure the computational power of a quantum computer based on factors like qubit number, connectivity, and error rates.
- ***Quantum Supremacy***—The hypothetical point where a quantum computer can solve a problem that is intractable for any classical computer. While not achieved yet, it would signal a major milestone.
- ***Quantum Neural Networks***—Proposals to use quantum computing for machine learning models and algorithms. May offer benefits like exponential speedups and capacity over classical neural networks. Still an active area of research.
- ***Qubit***—Unlike classical bits which can only be 0 or 1, qubits can be in a superposition of 0 and 1. This allows them to present more information and perform parallel computations on all possible inputs simultaneously.

M. Singh and B. S. Rawal, *Quantum Minds: Exploring Quantum Computing and AI Synergy*, https://doi.org/10.1007/978-981-95-2498-3

- ***Superposition***—A key property that gives qubits their power. For example, a qubit in an equal superposition of 0 and 1 can represent both values at the same time. This allows one qubit to essentially perform two computations in parallel.
  - A fundamental quantum phenomenon where a qubit can exist in multiple states at once. For example, a qubit in equal superposition of 0 and 1 can represent both 0 and 1 simultaneously. This allows a qubit to essentially perform computations on all possible inputs at the same time. The ability to leverage quantum superposition enables quantum speedups.
- ***Entanglement***—When two qubits are entangled, their states are linked, even if physically separated. Manipulating one qubit affects the other. This enables non-classical correlations and is key for quantum teleportation.
  - A phenomenon where two qubits become correlated in a way that is nonlocal. The quantum states of the qubits are linked, so manipulating one qubit instantaneously affects the state of the other, even if separated by large distances. Quantum entanglement leads to nonclassical correlations between measurements and is a key resource for quantum communication, computing, and cryptography.
- ***Measurement***—Observing a qubit causes its superposition state to probabilistically collapse to a single classical state. This destroys any superposition that existed before measuring.
- ***Wavefunction***—Represents the quantum state of a system through amplitudes of different configurations. The probabilities of measuring certain values can be derived from the wavefunction.
- ***Decoherence***—The loss of a qubit's quantum properties like superposition and entanglement due to interference from the external environment. This leads to errors in quantum computations. Keeping qubits isolated helps minimize decoherence. This destroys superpositions and entanglements, resulting in errors and noise. Suppressing decoherence remains a major challenge in building scalable quantum computers. Techniques like quantum error correction codes help overcome decoherence.
- **Quantum Complexity Theory**—The study of the resource requirements of quantum algorithms scale with input size. Quantum algorithms can achieve exponential speedup over classical algorithms by harnessing effects like quantum parallelism and entanglement. However, not all problems exhibit such speedups. Understanding quantum complexity helps identify practical applications.
- ***Reversibility***—Quantum gates and operations must be reversible to preserve information. This means no data can be erased in a quantum computation. Reversibility enables error correction mechanisms.
- ***Quantum Gates***—These gates manipulate qubits to perform quantum logic operations. The Hadamard gate creates superpositions, while gates like CNOT enable entanglement and interference.

- ***Quantum Circuit***—By linking quantum gates, more complex operations can be constructed. Quantum circuits operate on superpositions to carry out parallel operations.
- ***Quantum Algorithm***—Special algorithms like Shor's and Grover's are designed to take advantage of quantum properties like superposition and entanglement to achieve speedups over classical algorithms.
- ***Quantum Advantage***—For certain problems like factoring large numbers, quantum algorithms demonstrate exponential speedups over existing classical algorithms.
- ***Quantum Error Correction***—Decoherence tends to introduce errors into quantum states. Techniques like redundancy and syndrome measurements help detect and correct errors.
- ***Quantum Teleportation***—A process where the quantum state of one particle can be transmitted to another particle at a distant location. Enables transferring quantum information. Relies on entanglement and measurement.
- ***Quantum Key Distribution***—A method to securely distribute cryptographic keys using properties of quantum mechanics. Uses photon polarization and the no-cloning theorem to detect eavesdropping attempts.
- ***Quantum Cryptography***—Applications of quantum mechanics to perform cryptographic tasks like encryption, decryption, and signatures in a secure manner that is resistant to attacks. Offers security benefits over classical cryptography.
- ***Quantum Sensing***—Using quantum states and their unique properties to obtain very precise measurements of physical quantities like time, gravity, magnetism etc. Enables sensing smaller signals and forces.
- ***Quantum Repeaters***—Devices proposed to extend the range of quantum communication and entanglement distribution over long distances. Necessary for realizing global-scale quantum networks.
- ***Quantum Money***—Cryptographic schemes proposed to use quantum properties like entanglement to generate verifiable physical banknotes that are protected against counterfeiting.
- ***Quantum Debugging***—Techniques to detect and debug errors in quantum programs. Challenging compared to classical debugging since directly observing quantum states collapses superposition.
- ***Quantum Machine Learning***—An emerging field applying quantum computing to machine learning for potential benefits in speed, scalability, and modeling power over classical machine learning.
- ***Quantum Advantage***—The point where a quantum computer can solve a practical problem significantly faster than the best-known classical algorithm. While full quantum supremacy has not yet been achieved, some contrived examples and proofs-of-concept have demonstrated quantum advantage. True quantum advantage will lead to applications not possible on classical computers.

Glossary of Quantum Computing with Terminology

- *Qubit*—The basic unit of information in quantum computing. Unlike classical bits, qubits can represent a superposition of 0 and 1 states simultaneously due to quantum mechanical phenomena. Qubits enable massive parallelism.
- *Superposition*—The ability of a quantum system to exist in multiple states or values simultaneously. For example, a qubit in superposition could represent both 0 and 1 at the same time. This allows qubits to process information in parallel.
- *Entanglement*—A quantum phenomenon where two or more qubits become correlated in a way that actions performed on one qubit affects the others. This enables qubit connections with greater information density compared to classical systems.
- *Quantum Gate*—The quantum analogue of a logic gate in classical computing. Quantum gates manipulate qubit states similar to how logic gates act on classical bits. Common quantum gates include the Hadamard, CNOT, SWAP gates. Chaining quantum gates allows enacting quantum algorithms.
- *Quantum Circuit*—A sequence of interconnected quantum gates and operations that perform a quantum computation by transforming a set of input qubits through desired states to obtain the output. Quantum circuits manipulate qubit states to implement algorithms.
- *Quantum Algorithm*—A sequence of quantum operations like superposition, entanglement, interference and measurement applied to qubits in order to solve computational problems such as database search, factoring, or simulation. Key examples include Shor's algorithm, Grover's algorithm, Quantum Fourier Transform.
- *Quantum Machine Learning*—The interdisciplinary field applying quantum computing to machine learning tasks like supervised learning, unsupervised learning, and reinforcement learning. This allows leveraging quantum speedups for training ML models like neural networks.
- *Quantum Neural Networks*—Neural networks where some neurons have been replaced with quantum processing units to enable faster training by exploiting quantum parallelism for parameter updates.
- *Quantum Supremacy*—The stage where a quantum computer can solve a problem beyond the practical capabilities of even the most powerful classical supercomputers. First achieved by Google in 2019.
- *NISQ*—Noisy Intermediate-Scale Quantum computers - today's quantum processors with limited qubits, high error rates and lack of error correction. Algorithms must be tailored for NISQ limitations.
- *Quantum Volume*—A hardware-agnostic metric proposed by IBM to evaluate the performance of quantum computers based on factors like qubits, connectivity, gate fidelities, and compiler efficiency.
- *Quantum Advantage*—The point where quantum computers can solve valuable, real-world problems in areas like finance or drug discovery significantly faster than classical computers. Beyond just demonstration speedups.

- *Quantum Error Correction*—Techniques to detect errors like bit flips in qubit states using redundancy and then correct them. This will be essential for scaling to large fault-tolerant quantum computers.
- *Quantum Annealing*—Specialized quantum processors by companies like D-Wave that leverage quantum tunneling effects to accelerate solving optimization problems like planning or scheduling.
- *Quantum Generative Models*—Quantum machine learning models that learn to create synthetic samples like images, text, molecular structures by capturing key patterns in training data distributions.
- *Quantum Sensing*—Using quantum states like squeezed light or entangled photons to enable more precise measurement and imaging systems surpassing limits of classical techniques.
- *Quantum Cryptography*—Leveraging quantum properties like photon entanglement to enable secure communication channels that can detect tampering like quantum key distribution.

MIX
Papier aus verantwortungsvollen Quellen
Paper from responsible sources
FSC® C105338

If you have any concerns about our products,
you can contact us on
ProductSafety@springernature.com

In case Publisher is established outside the EU,
the EU authorized representative is:
Springer Nature Customer Service Center GmbH
Europaplatz 3, 69115 Heidelberg, Germany

Printed by Libri Plureos GmbH
in Hamburg, Germany